The Adirondack Park

The Adirondack Park

A Wildlands Quilt

BARBARA McMARTIN

SYRACUSE UNIVERSITY PRESS

Syracuse, New York 13244-5160

First Edition

99 00 01 02 03 04 6 5 4 3 2 1

This book is published with the assistance of a grant from the John Ben Snow Foundation.

The paper used in this publication meets the minimum requirements of American National Standard for Information Sciences—Permanence of Paper for Printed Library Materials, ANSI Z39.48-1984

LIBRARY OF CONGRESS CATALOGING-IN-PUBLICATION DATA

McMartin, Barbara.
The Adirondack Park : a wildlands quilt / Barbara McMartin.—1st ed.
p. cm.
ISBN 0-8156-0567-6 (pbk. : alk. paper)
1. Adirondack Park (N.Y.)—Description and travel. 2. Adirondack Park (N.Y.)—Guidebooks. 3. Adirondack Park (N.Y.)—Pictorial works. I. Title.
F127.A2M36 1999
917.47'50443—dc21 98-49278

PRINTED IN CANADA

MAPS BY W. ALEC REID
All maps drawn to the same scale (1 inch = approx. 5.6 miles)

FRONTISPIECE: *Lake Lila Primitive Area and Little Tupper (see pages 50–51). The most important state acquisition in 1998 was Little Tupper Lake and land surrounding it in the Whitney Tract. The lake is dotted with small, rocky islands and peninsulas topped with dark evergreens.* Photograph courtesy of the author.

Contents

Pharaoh Lake Wilderness (see pp. 2–3). Views from hard-to-reach cliffs are the rewards of long hikes and bushwhacks. The view toward Brant Lake and the Brothers Mountains from Barton High Cliffs is one of the best of those rewards. Photograph courtesy of Chuck Bennett.

Piecing the Quilt

A crazy quilt—a patchwork quilt without a design.

Look at a crazy quilt and marvel at the way odd-shaped pieces have been fitted together—here a rectangle, there a triangle, trapezoid, or square. Each piece has a different shape, but the whole lies flat, like a map. Each piece is seamed to the next, and the seams are accentuated by different embroidered chains. Silks are juxtaposed with velvets, prints with plushes, stripes with flowered patterns. Some fabrics are repeated, but there is no plan to the repeats, no overriding theme. It is as if the quilt just grew, piece by piece.

Even without a plan or design, there is a beauty to the construction, a synergy of texture and color, and the vibrancy and glow of many different pieces working together to create a pleasing work of art.

Our foremothers created crazy quilts to treasure, and to warm their families; our forefathers created a crazy quilt of land parcels that is today's Adirondack Park.

The analogy is appropriate, for the two creations have much more in common than is readily apparent. Certainly the quilt and the Park were formed of different shapes stitched together. Just as a treasured quilt was created from odds and ends, so the Adirondacks became a treasure with many disparate parts.

The choicest pieces in the quilters' crazy pattern were remnants of the finest garments, the hardest to throw away. Some were so worn that even though they were treasured, they were ready to be discarded. The parcels that make up the Adirondack Park, a mixture of private lands and the public lands called the Forest Preserve, are treasured, but most of the Forest Preserve components were at one time ready to be thrown away. Although it did not seem so when the Forest Preserve parcels were being assembled, the analogy of cast-off parts certainly applies. The choicest pieces of forest today are those that at one time had the least value, that were the first to be abandoned. They had been abandoned because their merchantable timber had been logged. Because there was no other means of transportation, lumbermen were restricted to floating logs downstream to mills outside the region. Because only softwoods would float, only softwoods were considered merchantable; and when the pine, spruce, and hemlock trees were gone, the tracts from which they had been harvested were considered worthless by the lumbermen, even though most still contained stands of virgin hardwoods. The value of the land with its rich hardwood stands was considerably less than the taxes assessed.

The quilters acquired a few choice pieces of silk and velvet to give consistency to the random patterns—inspirations for each quilt's color scheme. The state acquired a few choice parcels that had never been logged—swaths of great forest whose beauty inspired all acquisitions with the promise of preserved forest. Interspersed as well are the many parcels that remain in private hands, the basis of the modern forest industry. These, too, became part of the mosaic that is the Adirondack Park.

The quilters joined their pieces with varied stitches. The patches of industrial lands and Forest Preserve appear stitched together along jagged boundaries that most often followed rivers or narrow valleys with their chains of settlements that dot the Park.

The creation of the Forest Preserve illuminates the story of how the lands nobody wanted were pieced together to become the most sought-after and protected lands in the east.

In 1885, New York State set aside the then public lands within a blue line drawn around the Adirondack region and called them the Forest Preserve. That line already encompassed nearly a half million acres of virgin lands or lands that had only been harvested for scattered stands of softwoods. In the next two decades, more lands were abandoned by their owners and added to the Forest Preserve. The state began to see how valuable that landed treasure was, and it voted to buy still more parcels, until today more than two and one-half million acres of public lands are the most valued fabrics of the Adirondack quilt.

As the Forest Preserve has grown, a succession of political decisions has determined the way the people of New York have dealt with these public lands. The most important was the 1894 constitutional amendment that placed the Forest Preserve clause in the state's constitution. The most problematic may have been the management guidelines established in 1972 by the Adirondack Park Agency Act and the subsequent development of the Adirondack Park State Land Master Plan (APSLMP). That plan described the units of Forest Preserve found in the Park and prescribed specified management goals for each.

The various elements of the Adirondack Forest Preserve are remarkably dissimilar. They differ in their forest cover, their lakes and mountains, their natural history, geography, and human history. Their geographical origins, waterways, and forest cover determined the history of their settlements or lack of settlements. The history of their use by humans determines in part their present natural conditions. To begin to understand the differences between the patches of Forest Preserve, it is first necessary to understand how the patches of state land were cut into the odd shapes that have inspired the notion that the Forest Preserve has a pattern resembling a crazy quilt.

Naming and designating the management goals for various regions of the Forest Preserve came about in a haphazard and circuitous way, only describable in terms of the historical record. Geography played a role—roads were first placed in river valleys along the levelest courses, thus leaving the high lands outlined with the network of roads. Because rivers and streams were the only mode of transportation for logs, and because trees were the region's only export, with the exception of iron from the eastern Adirondacks, this further emphasized the way Adirondack regions were outlined by their encircling rivers.

These regions varied greatly in size. Among the largest was the area we now know as the Silver Lake Wilderness. Its boundaries are roughly the West and Main Branches of the Sacandaga River and the chain of lakes from Piseco to Speculator. Just as in many other regions of the Park, no real through road, at least one that made economic sense, ever penetrated or traversed its wild interior. This vast dome of wild lands was considered valuable only for its forests, and with the exception of areas near its river boundaries, its forests were very difficult to harvest and often found to be devoid of stands of the much-sought-after spruce.

Having a river did not necessarily mean that there would be a road: witness part of the West Branch of the Sacandaga from Piseco Outlet to Whitehouse or the Hudson River below Newcomb. Some river valleys were just too rugged for parallel roads.

What the network of roads did do, however, was outline regions where access to the interior was difficult. The cores of these regions were inhabited only by a few isolated sportsmens' or loggers' camps. As logging declined, as it did after 1910,[1] the interior regions generally became less and less frequently visited. The one exception to this was the High Peaks region, where guides had begun taking explorers in the early part of the nineteenth century and where, by the second half of that century, guides had cut trails up most of the higher mountains. By the twentieth century, trails laced most of the High Peaks and hiking organizations were formed to promote and maintain them.

Except for the extraordinary interest in the High Peaks, the rather amorphous public lands in the Adirondack region remained known simply as Forest Preserve lands until after World War II; the postwar time brought demands to accommodate motorized vehicles, particularly snowmobiles. It is peculiar that this period coincided with the rise of wilderness concerns throughout the nation's parks. Wild places were becoming so popular that people glimpsed for the first time the possibility that they could be overused, and the access permitted by motorized vehicles made that all too real a possibility.

In the late 1960s, the Temporary Study Commission on the Adirondacks changed the way New Yorkers viewed their Forest Preserve. Like all political

1. Barbara McMartin, *The Great Forest of the Adirondacks* (Utica, N.Y.: North Country Books, 1994).

decisions, that commission's work, which resulted in the Adirondack Park Agency Act, was a series of compromises. The compromise concerning the Forest Preserve called for splitting it into two principal categories, Wilderness and Wild Forest, with any form of motorized access permitted only in Wild Forests. In essence, only snowmobiles were allowed there, except on existing roads. Let's put off the discussion of the ramifications of this for a moment so we can conclude the discussion of the naming and designating of the various areas.

The Temporary Commission roughly drew boundaries around the more or less contiguous patches of state land. Some of these were well defined, some were not. Most of the boundaries followed the already established network of roads and river corridors. Some patches had penetrating roads, others did not. They ranged in size from several hundred thousand acres to less than a thousand acres.

The intent was to set aside the largest and most well-defined regions, those that were the least chopped up by penetrating roads, those that had the fewest private land inholdings (parcels surrounded by state land), and those that could be most easily protected. These would be called "Wilderness," and they were further expected to have other attributes set forth in the 1964 Federal Wilderness guidelines—a place where man was only a visitor, where his works were not obvious, where solitary and unconfined recreation was possible, where nature dominated.

The Adirondack Park Agency Act differs from the Federal Wilderness Act with respect to minimum size of a Wilderness area. In the Adirondacks they must be at least 10,000 acres; the federal act permits areas as small as 5,000 acres to be called Wilderness. The most significant difference between the two is that the Adirondack Park Agency Act defines the category "Wild Forest" for all the Forest Preserve areas that do not meet the Wilderness definition. These Wild Forests account for more than half of the Forest Preserve.

The regions were given names that derived from long tradition. When the Adirondack region was first mapped, it was divided into quadrangles, each of which was given the name of its most distinguishing feature, usually a prominent lake, though sometimes a mountain. Few regions extended beyond the range of their central quadrangles. The names of those quadrangles had already been given to some areas by map makers or tradition; now those names were adopted for the newly defined political entities.

Just to confuse matters, this naming is but one way of describing a part of the Adirondacks. Another address system derives from the first surveyors, who named large tracts and patents after their first owners. Each of these patents or tracts was further divided into townships, usually numbered, but sometimes named, and again further divided into numbered lots. The map on page 1 shows the principal divisions discussed in this book.[2]

Why were some regions called wilderness and others called wild forest? In spite of what purported to be careful natural resource studies of the different areas, the decisions about their designations were often based on political considerations. For example, in the southeastern Adirondacks the region bounded by the East and Main Branches of the Sacandaga, the Sacandaga Reservoir, and the Hudson is one of the largest naturally defined areas in the Adirondacks. Two roads thrust toward the interior from the east, one passing Harrisburg, the other reaching Baldwin Springs. There was one private inholding and significant tracts of private land in the southern reaches. This region was designated the Wilcox Lake Wild Forest.

Compare this designation with that of some of the wilderness areas. The Silver Lake area has private inholdings and a road that plunges into its heart from Wells. The High Peaks has roads from the north toward Heart Lake and from the south to the huge private lands around Tahawus. By all measures, the Wilcox Lake area deserved to be designated wilderness. Oddly enough, the designation means little. All of the Forest Preserve is intrinsically wilderness.

Responding to the way recreationists, other than hunters and snowmo-

2. See McMartin, *Great Forest,* 15–19, for more details about forest history.

bilers, ignored these lands, public policy also ignored the wild forest areas. As a result, most of those lands remain essentially wilderness regions. They have not attracted the hiking, skiing, wild-land-seeking portion of outdoors visitors who respond to the cachet of the word "wilderness." Planners from the time of the Temporary Study Commission have concentrated their recreational efforts on those wilderness regions; in fact, with two exceptions, most trails for recreationists have been concentrated in the High Peaks Wilderness.

A wilderness is an area with a primeval character, dominated by nature, without evidence of significant human activity, offering opportunities for solitude. Guidelines for wilderness thus specify large size, absence of private inholdings, few or no incursions by roads, and natural boundaries that make it possible to define and protect the area. A wild forest is defined as an area that permits a higher degree of human use than wilderness and frequently lacks the sense of remoteness of a wilderness. Snowmobiles are permitted in wild forest areas; in fact, they are the only motorized vehicles permitted. In theory, natural conditions should prevail in wilderness; recreation should be important in wild forest. That has not occurred, however.

The difference between wild forest and wilderness was blurred in practice when the Forest Preserve was divided into these two different types of areas. In reality, wilderness and wild forest areas form a continuum of wildness that is best understood by considering them all as discrete parts of the Forest Preserve.

The blurring of the distinction between wilderness and wild forest in practice was partly political and partly because of shortfalls in the original studies that were made to differentiate the areas. Naturalists looked at some of the rivers, lakes, and forests and analyzed them without placing them in a historical context. Indeed, the history of the forests was little understood when the APSLMP was prepared. Past fires were noted, but different types of logging were not appreciated. It is in the context of forest cover that the similarities of wilderness and wild forest areas become most apparent, but that is to be expected—they are all parts of the forest preserved.

The trails in the wild forests mostly follow old logging or tote roads, roads in flat areas that are often wet. They make great snowmobile trails. They are not the best-designed hiking trails. Except for routes to fire towers, very few mountains in wild forest areas have hiking trails. The wild forest areas have been neglected, overlooked, and allowed to become even wilder.

This book is planned to right that imbalance. It will not only show how wonderful and diverse are our wilderness areas, but how little they differ from our wild forests. It will return our perspective to all the patches of Forest Preserve and introduce the reader to the essence of each patch, the characteristics that differentiate each one. In delineating a menu of recreational activities, it will show how people might participate in recreational activities in the setting they prefer. It will mention the recreational spectrum available and hint at the inadequacies of recreation planning that has failed to develop that potential. Finally, it will show the nonhiking public, including those with disabilities, ways to view these regions as gems of the Forest Preserve.

This work is the culmination of many of my writings. My *Discover* guides[3] showed hikers how to find most of the secrets of all the regions, treating all parts of the Forest Preserve equally. When I wrote on the Forest Preserve for the Adirondack Park Agency,[4] I attempted to encapsulate that which distinguishes each region; but I had to write that booklet within the space and outline defined by the agency. Similarly, when I wrote about the recreational potential of wild forests[5] for the Adirondack Council, I was limited by that organization's mission and was unable to extend my concern for

3. A series of eleven regional guides that cover the entire park, published by Lake View Press.

4. State of New York, Adirondack Park Agency, *Citizen's Guide to the Adirondack Forest Preserve,* designed and written by Barbara McMartin, 1985.

5. Barbara McMartin, *Realizing the Potential of Adirondack Wild Forest,* vol. 3 of *2020 Vision—Fulfilling the Promise of the Adirondack Park,* (Elizabethtown, N.Y.: The Adirondack Council, 1990).

recreational opportunities to council's vision for wilderness.

This book is my own work, although Betsy Folwell, the editor of *Adirondack Life,* has offered several important suggestions, and Mary MacKenzie, historian of Lake Placid and North Elba, provided me with details of forest history in three northern wilderness areas. The book gives my views of what constitutes the parts of the Forest Preserve, combined with a few suggestions for ways to enhance their worth. It is an attempt to distill the natural and human history of each region, to show the value of each segment of the Forest Preserve. It emphasizes their "preserved" qualities by showing that they are all essentially wilderness. It invites the reader to become acquainted with all our wildernesses, or more properly, all parts of our Forest Preserve.

My descriptions are as varied as the areas themselves. The short essays focus on the inherent characteristics—landscape and forests—of each area, ignoring changing seasons and the ephemeral and seldom seen birds and animals. Accompanying each short essay is a suggestion of places to enjoy views and visits or short walks so you can sample the values I want to share.

The Adirondack Park

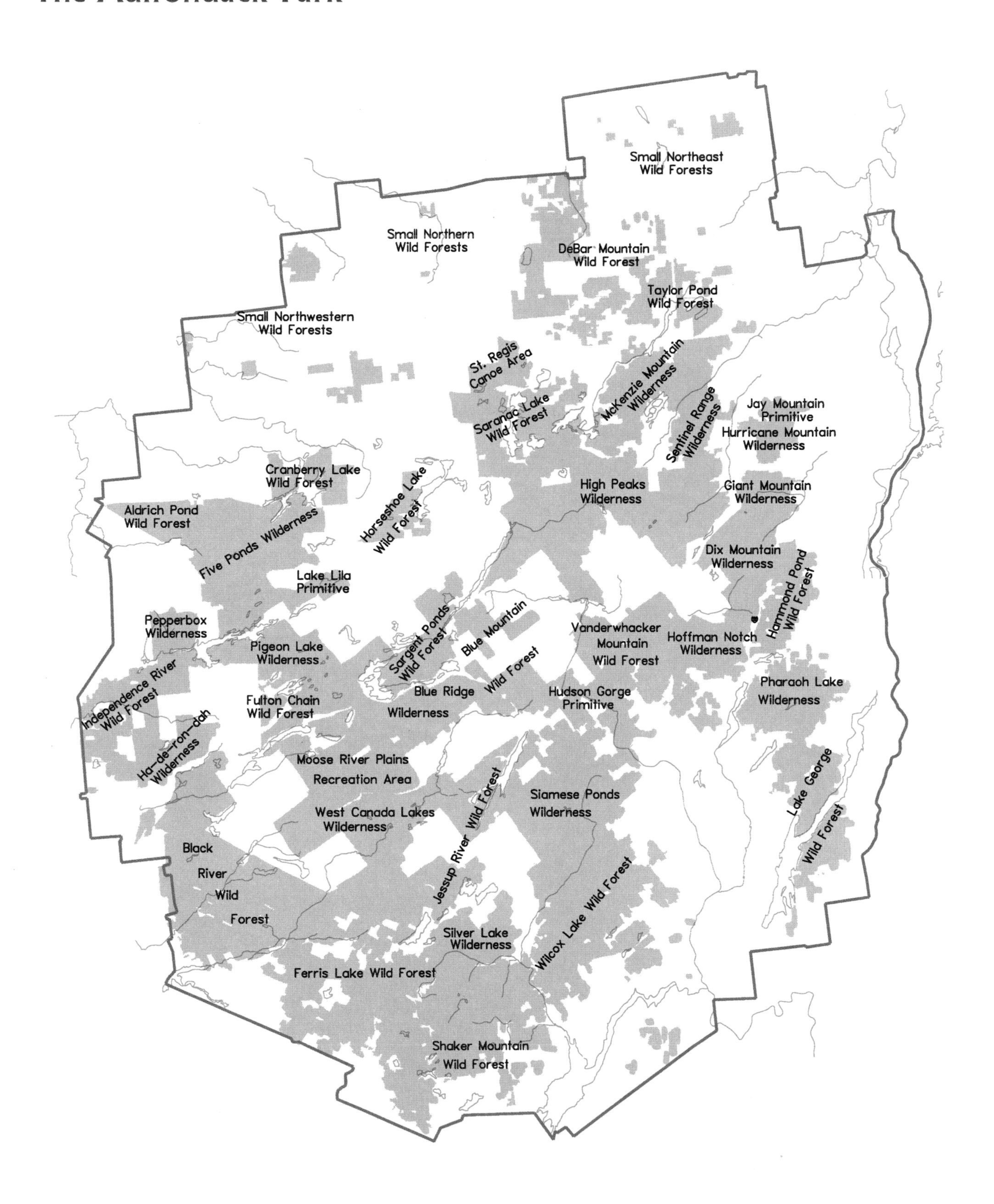

1 Pharaoh Lake Wilderness

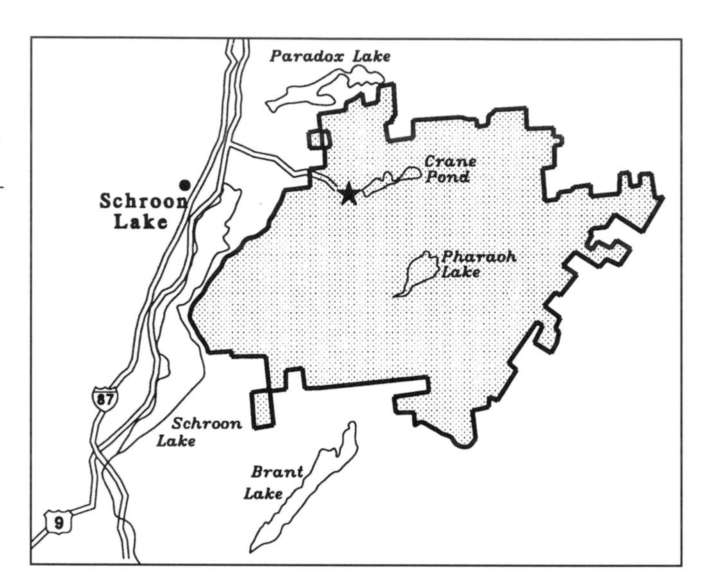

The Pharaoh Lake Wilderness is a planned wilderness of 46,000 acres. It was acquired in the early years of the twentieth century by purchase with some of the first funds appropriated by the state for land acquisition.

Although it represents one of the most important purchases ever made for the Forest Preserve, the Pharaoh Lake Wilderness typifies the problems the state has encountered from the beginning of its land reacquisition efforts. In the early 1900s, when the state had money to purchase land (one million dollars was appropriated in 1897, one-half million in 1898, and smaller amounts in the next few years), there were numerous incidences in which individuals purchased land ahead of the state, held it briefly, and then sold it to the state for enormous profits, often twice what had been paid. In the 1890s and the first decade of the twentieth century, land values escalated, merchantable timber became scarce, and millionaires and private clubs competed with the state and with lumbermen for pristine tracts of land.

A number of entrepreneurs, many of them lawyers, appeared on the scene. They sought to profit from this scramble for land. The most successful and cleverest man at this game was one George N. Ostrander, who purchased about half of the future Pharaoh Lake Wilderness from the Pickard estate. Heirs of that estate were in England, and the state was negotiating with them for the property. Ostrander was able to move more quickly. Not only did he double his investment and realize a profit of $70,000, he retained the strip of land on the east shore of Schroon Lake. This he sold to private individuals, generating as much profit as did the sale of the rest of the tract. Shortly after, the state purchased another large chunk of the future wilderness from the Raquette Falls Lumber Company that included Crane Pond in its holdings. Ostrander was secretary and stockholder of that lumber company, which had purchased the land in 1903 for two dollars an acre. The state purchased it in 1908 for $7.25 an acre.

Almost all of the region had been logged, some of it before 1850 for spruce, much of it in the 1860s and 1870s for hemlock bark for the tanneries near Schroon and Brant Lakes. Evergreens always thrived in the region, and most notably, pines flourished on the sandy flats between mountains.

Fires were especially severe, burning the tops of almost all the mountains, big and small. From the giants Pharaoh and Treadway down to the small hills like Ragged and Potter and Bear, many summits were denuded by fire. The summits that remain bare are among the tract's most enduring charms. The variety of views is endless. Lower-elevation forests, however, soon recovered from fires and logging, and the forest cloaks that shroud the deep valleys have almost regained their original grandeur.

But the most attractive feature of the region is the juxtaposition of the myriad lakes and ponds with their bordering sharp rock outcrops, small cliffs, and ledges. Reflections of rock and forest in the quiet waters make each watery vista a picture to savor. Nowhere else are so many delightful shores connected with such a fine network of trails.

The ponds' names reflect the fact that they were originated by real visi-

In all the Pharaoh Lakes Wilderness there are three dozen ponds, each with a distincitive shoreline, but there is none so special as the natural sculpture garden with its own reflecting pool at the head of Pharaoh Lake. Photograph courtesy of Chuck Bennett.

tors, not surveyors with their limited vocabulary. For the armchair visitor, just reading the list of places takes you into a fanciful realm: Tub Mill Marsh; Grizzle Ocean; Oxshoe, Horseshoe, and Crab Ponds; and Lilypad, Bear, Goose, and Crane Ponds.

Despite the expense at the time the land and ponds were purchased, no finer acquisition could have been made.

Views and Visits

The road leading to Crane Pond is officially closed, after a long controversy. The closure was needed to protect the wilderness, and the access road now provides an easy walk through beautiful forest stands and along a small gorge to a lake that is much quieter now.

2 Lake George Wild Forest

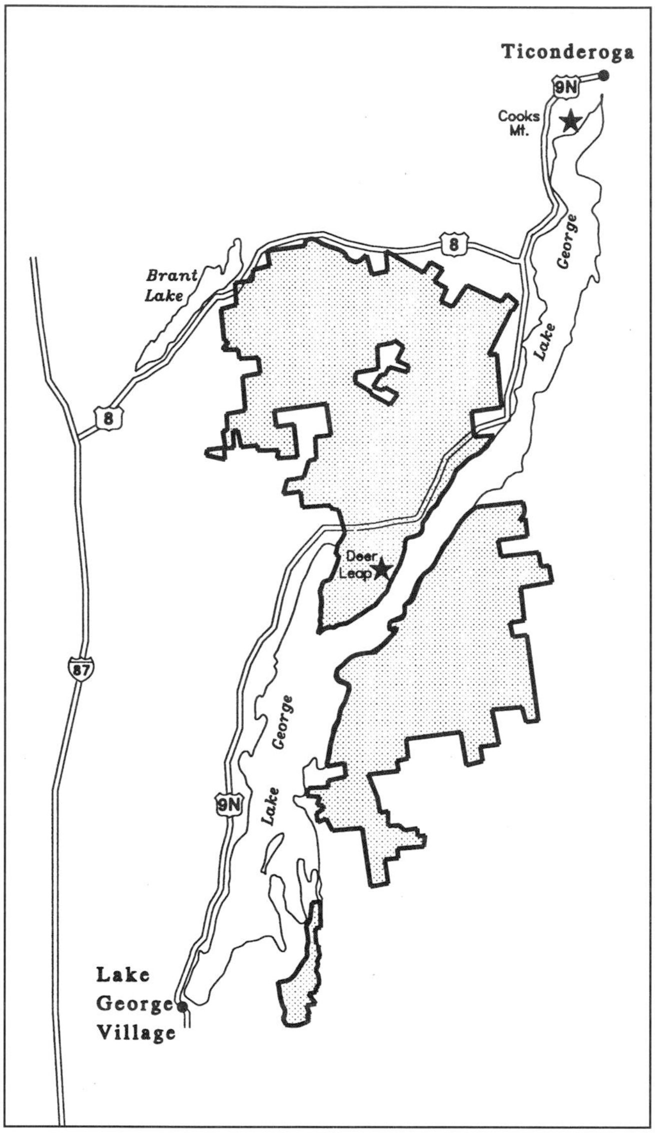

Think of Lake George and you think of pure waters dotted with islands and shores lined by steep mountains. These mountains and their trails make this a wild recreation area that contrasts vividly with the settled region and tourist attractions at the lake's southern end. This wild forest is immense; its 120,000 acres protect vast tracts along Lake George's thirty-mile length.

Imagine the scenes these mountains have observed—Indian raiders, armies on snowshoes, fleets of bateaux, the siege before the fortress of Fort William Henry, elegant hotels, and cruise ships.

Lake George owes some of its majestic forests and wooded shoreline to private landowners. Over seven thousand acres on the southeastern shore were purchased by George Knapp in 1894. Almost all of the former Knapp estate is now Forest Preserve, and its eighty miles of carriage roads are now hiking and horse trails. The tract had been logged before Knapp acquired it, but it has never been logged again, so the stands of hemlocks are as magnificent as they were in 1607 when the first white man, a priest captured by Indians, visited the lake.

Public land stretches north from the former Knapp estate to Black Mountain, then discontinuously north through the newly acquired shoreline of the Morgan property and on to Spruce Mountain with its great views.

A long peninsula thrusts south parallel to the western shore. Tongue Mountain is a ridge with a series of open peaks and extraordinary views. Farther north, Rogers Rock and Cooks Mountain complete the western shore's line of sentinel peaks with views.

Old roads and isolated ponds complete the wild forest area northwest of the Tongue—this tract reaches all the way to Brant Lake and encompasses a former fire-tower mountain that needs to rejoin the list of recreational peaks.

Views and Visits

Most of the trails are long. Deer Leap, with its spectacular view down Lake George, or Cooks Mountain near Ticonderoga, with its commanding view of the north end of the lake, are among the shorter walks.

Lake George is truly the Adirondack Park's most photogenic lake. Here the panorama of the Tongue Mountain Range and Black and Buck Mountains is enhanced by clouds and the lake's blue waters.

Photograph © 1998 by Carl E. Heilman II.

Streaks of winter sun dazzle ice-bound Lake George. It is more difficult to reach mountain heights, here Pilot Knob, in winter, but the frozen image is incomparable.

Photograph courtesy of Chuck Bennett.

3 Wilcox Lake Wild Forest

Wilcox Lake Wild Forest is so big and varied it could be the most important recreational area in the Park. It probably should have been designated wilderness; most of it really is.

The small mountains in the range that stretches along the eastern boundary yielded their forests very early to settlers and loggers, who floated softwoods to the mills near Glens Falls. Many of the summits were burned in the turn-of-the century fires. From north to south the range includes Huckleberry Mountain with its red pine ridge; Crane Mountain with its fields of lichens and blueberries beneath scattered pines; Bear Pen with its dizzying cliff-top vista across the sand plains that fill the valleys of Madison and Stony Creeks; Baldhead and Moose with its views of Crane Mountain as well as the marvelous birch-filled valley that separates the two peaks; and the long ridge of West Mountain that ends in Hadley Summit with its fire tower and a distant panorama that stretches from the high peaks of the Catskills to the high peaks of the Adirondacks. Only Crane and Hadley have trails; the others deserve trails.

The East Branch of the Sacandaga forms the northwestern boundary of the wild forest, and since N.Y. 8 is south of that river, you would expect many trails to lead south from it. A few do; snowmobile trails and fishermen's routes reach Fish and Kibby Ponds. The old road along Stewart Creek to Oregon was one of the first roads north from

Crane Mountain rises above the shore of its eponymous pond, just one of the great views of rock and trees and water from this great massif. Such open rock summits are typical of this southeastern Adirondack wild forest. Photograph courtesy of the author.

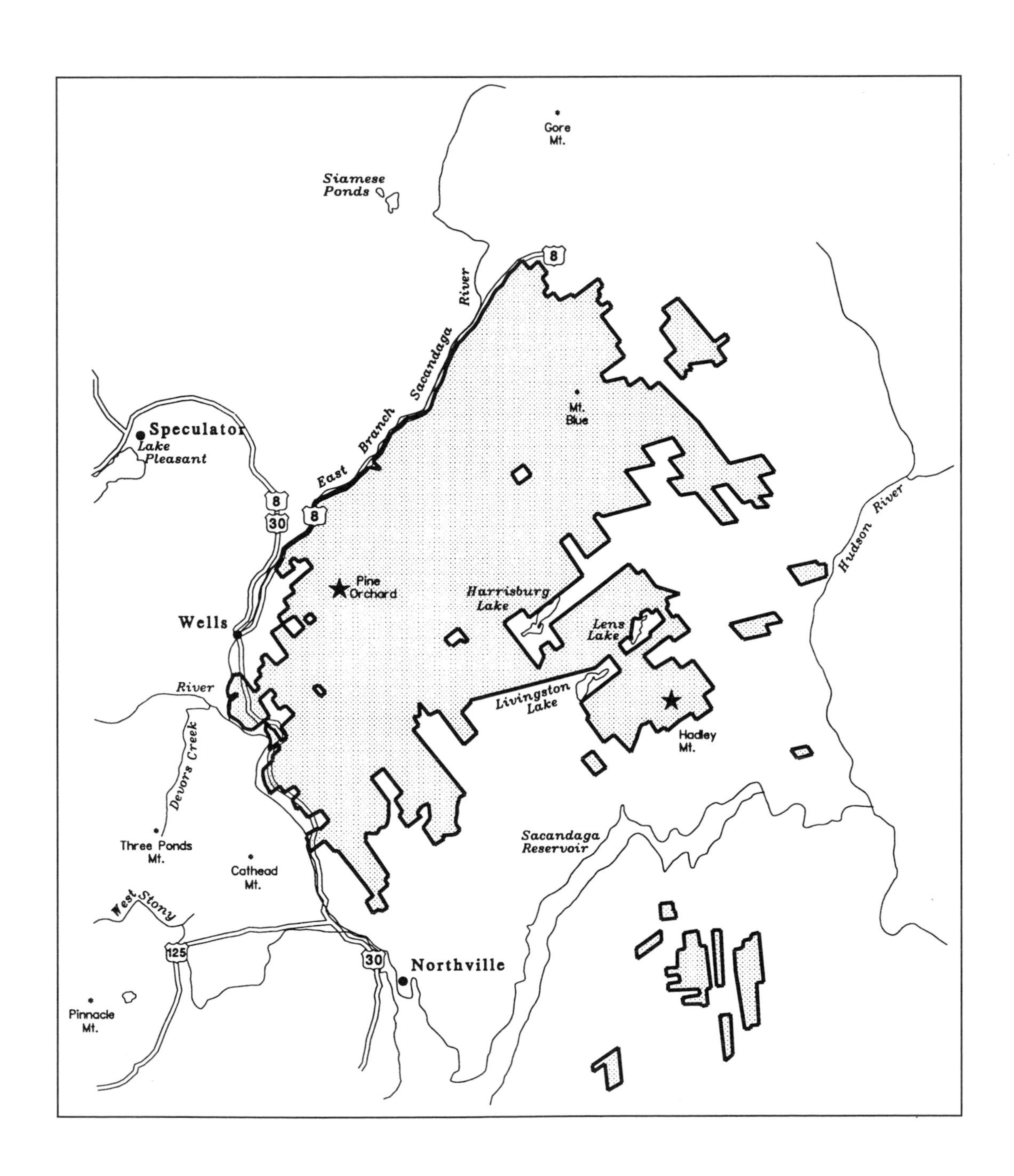
Gore Mt.
Siamese Ponds
8
River
Sacandaga
Branch
East
Mt. Blue
Speculator
Lake Pleasant
8
30
8
Pine Orchard
Harrisburg Lake
Lens Lake
Hudson River
Wells
River
Livingston Lake
Devors Creek
Hadley Mt.
Sacandaga Reservoir
Three Ponds Mt.
Cathead Mt.
West Stony
125
30
Northville
Pinnacle Mt.

Stony Creek. The road south from Georgia Creek to Wells climbs through Pine Orchard, a knob covered with giant pines that have been growing since an 1815 hurricane. The old road from Willis Lake to Wilcox Lake winds through a spectacular valley bordered on the north by hardwood slopes that were never logged.

A dense network of trails traverses all the region's valleys. Most of the valleys that were logged for softwoods have been state land since just before the turn of the century. What you may never notice as you travel along these old roads, now marked for snowmobiles, are the magnificent hardwoods that grace the slopes above these valleys. Virgin maple, beech, and yellow birch stands grow on the slopes of Georgia, Smith, and New Lake Mountains and the upper slopes of Spruce, Steve Bigle, and Cattle Mountains.

One mountain stands out on the western border of the wild forest—Moose Mountain rises above the Sacandaga Valley south of Wells. On its southern slope is a protected cove sheltered by its summit-ridge cliffs. That cove contains the most wonderful stand of maples, birch and ash—it has never been logged. The majestic maples south of Griffin constituted a prime sugarbush.

The western hills are also trailless, yet the natural views on Rand, Moose, and Murphy Mountains call out for trails.

Waterfalls on Tenant Creek, a chain of lakes (Murphy, Middle, and Bennett), the trail along Stony Creek, the marshes in Madison Creek Flow, and the deep, fern-filled valley of Jimmy Creek are all special places as equally deserving of trails as the wild forest's mountains. I am lucky because I have visited all these places. Few people have, and yet as a wild forest, these places ought to be accessible to all outdoors people, especially family groups.

Views and Visits

With a guidebook, almost everyone can enjoy the two-mile walk to Pine Orchard. Hadley Mountain Trail is also a relatively easy hike.

Three lovely falls grace Tenant Creek, but many others can be found in the streams of the Wilcox Lake Wild Forest.

Photograph courtesy of the author.

4 Siamese Ponds Wilderness

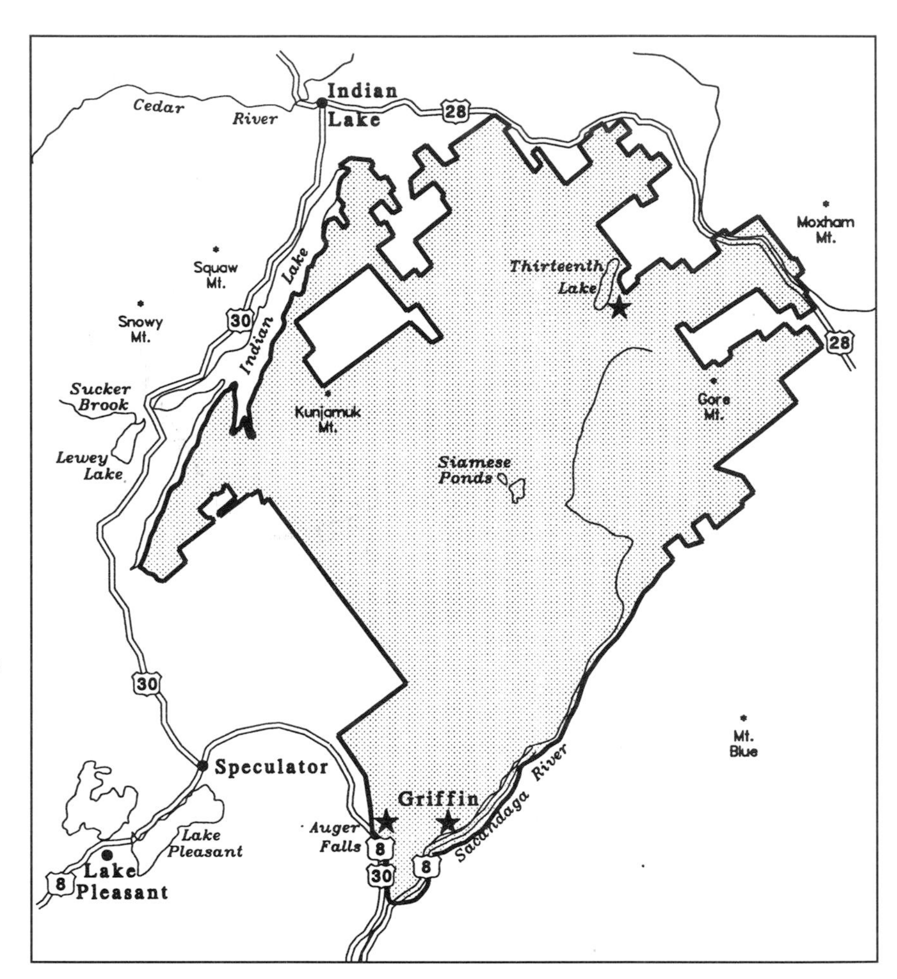

This wilderness has much more to distinguish it than its eponymous ponds. Even though the northeast corner of the wilderness was settled in the early 1800s, and the railroad reached North Creek in 1871, its interior never attracted settlement. Perhaps the geographical characteristics of its well-defined boundary features account for its seclusion. The long valleys of the Indian and Jessup Rivers, later dammed to create Indian Lake, form a barrier on the west. The mountain chain beginning with the garnet mountains (Gore and Ruby) in the north stretches south through Height-of-Land Mountain to Eleventh Mountain. This chain formed a barrier that stopped expansion from the Hudson River Valley, which lies to the east of those mountains. Puffer and Bullhead Mountains stopped invasions from the north.

Two very old and remarkably level valley routes severed the region from north to south. The one on the west preceded the route of the modern road (N.Y. 30) along Indian Lake by more than a century. It started along the valley of the Kunjamuk, passing the Rhinelander estate, which was built at Elm Lake in 1815 and abandoned shortly afterwards. North of the Kunjamuk Valley, the road crossed to the Kings Flow drainage and continued toward Indian Lake village.

The route on the east avoided the gorge on the East Branch by climbing up and over the steep shoulder of Eleventh Mountain. It continued north along the East Branch until it could cross to the Thirteenth Lake drainage. There, beyond the old garnet mine settlements near Thirteenth Lake, the road began to descend steeply as it followed Thirteenth Lake Outlet down to the Hudson.

Only the steep section of road, which originally served mining settlements, survives on the east. The rest, from Old Farm Clearing south along the East Branch to N.Y. 8 is a wonderful trail. On the west, the road as far south of Indian Lake as Kings Flow survives as a road. The southern part of this road would probably also have disappeared were it not that it remains the backbone of the logging roads in International Paper Company's diamond-shaped tract, which lies at the southwest corner of the wilderness. The middle portion of this old road is fast disappearing.

Auger Falls is a wild place, almost too dangerous to photograph in high water when the flume is at its most magnificent.

Photograph courtesy of Chuck Bennett.

Some of the original roads in the north leading to John Pond and south to Puffer and east to Thirteenth Lake survive as cross-country ski trails.

The East Branch of the Sacandaga flows through a deep fault valley bordered with sharp hills that rise above the river to the north. The river, with only one bridge (at Griffin), is an effective barrier on the southeastern side of the wilderness. Turn-of-the-century fires raged across those mountains—Buckhorn, Corner, and Black Mountain and Shanty Cliffs on the Blue Hills all have patches of bare rock that attest to the severity of the fires. Flood dams for logging were built upstream on every creek that flows south into the East Branch. Routes along these streams, which cleave the mountain range, offered the only level routes for tanbarkers. Only a few fishermen and hunters follow these routes today.

One of the Adirondacks geological wonders—Chimney Mountain—lies in the northwest corner of the wilderness. When the last glacier receded, the top of the mountain was split, exposing layers of Grenville-era rock and the chimney. The release of pressure also opened caves beneath the cleft valley and the walls above.

Views and Visits

Walk down the road to Griffin to the iron bridge above the East Branch, enjoy the waterfalls below the bridge, or discover the way the nearby forest conceals the long-abandoned tannery settlement. Marvel at the way the forest is hiding signs of the buildings that lined the river here; wonder at the way the forest conceals the spirits of those settlers.

Alternatively, visit the picnic spot on the Main Branch of the Sacandaga, near Auger Falls; canoe Thirteenth Lake through a fault valley that pierces the northwestern corner of the wilderness; or just drive along N.Y. 8 and enjoy seeing the East Branch and the mountains to its north.

Chimney Mountain beckons picknickers, cavers, geologists, and just plain explorers to this geological wonder in the Siamese Ponds Wilderness. Photograph courtesy of Chuck Bennett.

5 Silver Lake Wilderness

The West Branch of the Sacandaga River defines part of the boundary of the Silver Lake Wilderness, but, more important, it coils protectively around its remote center. In part, the spiraling river and the ruggedness of the enclosed terrain account for the fact that the inner core of this third largest and most undeveloped of all our wilderness areas has been shielded from incursions. The real reason for its security, however, has less to do with its geography or active preservation and more to do with what it was never able to offer. The region's forests of magnificent northern hardwoods lightly mixed with scattered spruce and balsam are some of the most noble in the Forest Preserve. But they were generally a disappointment to those who first claimed the forest.

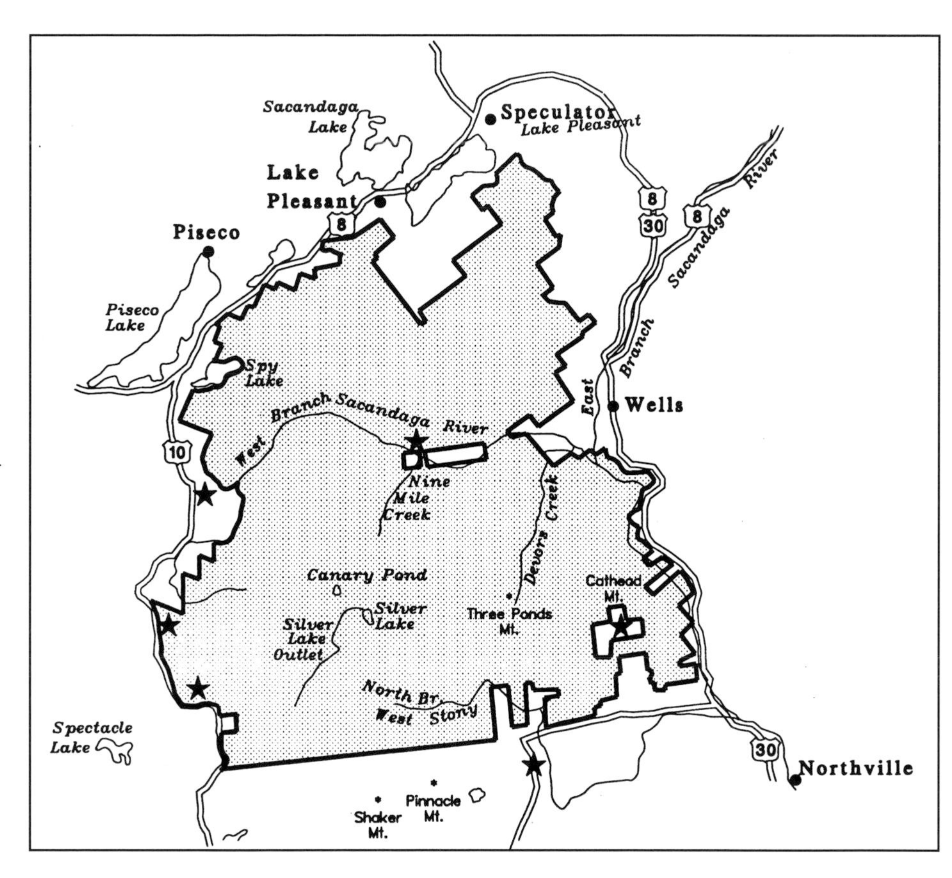

The trapezoidal boundaries of Benson Tract fill out much of this wilderness area. Benson Tract was surveyed in 1795, and a small part of its southeastern corner was settled around 1800. Lumbermen reached out along logging roads to the north and west and along the West Branch itself, but they quickly learned that Benson's forests and those of Chase's Patent and Glen, Bleecker, and Lansing Patent to the south had little merchantable timber. Pockets of spruce filled small valleys and lined the river banks, but unwanted beech, maple, and yellow birch covered most of the mountain slopes. Hemlocks filled Groff Creek Valley and the West Creek Valley but grew only sparsely elsewhere.

In the south, logging roads thrust north to the notch between Wallace and Three Ponds Mountains and on Three Ponds Mountain itself. A sawmill was erected on Abner Brook. The dense, dark stands of hemlocks that lined Groff Creek valley beckoned tanners from Wells. Stands of tall, straight hemlocks grace the steep slopes above the creek as if they had always been there. A road through the Groff Creek valley once stretched all the way to a logging camp at Devorse Creek. It passed through such a deep, sheer-sided gorge that a flying trestle had to be build so horses could pull logs from the vicinity of the camp.

Improbably, a dam was built at the headwaters of Nine Mile Creek. Helldevil Dam, on the north slopes of Three Ponds Mountain, added a surge of water to float logs down the West Branch from the slopes of nearby Whitehouse. The West River Road brought loggers from Wells to the spruce- and hemlock-covered borders of the river all the way to the confluence with Piseco Outlet.

Loggers approached the region of the West Branch's three headwater ponds, Silver Lake, Canary Pond, and Meco Lake, which lie west of Three Ponds Mountain. Approaching from the southwest, loggers built flood dams at North Branch Reservoir on the North Branch and along Silver Lake Outlet. They harvested spruce logs and sent them down those tributaries to the West Branch. There the logs floated north along the now sluggish river through the sand-flats created by glacial Lake Piseco. At the north end of the flats, the river turns east and, joined by Piseco Outlet, drops through a deep gorge where waterfalls and cataracts appear to make the river an impossible carrier of logs.

As quickly as the logging camps and dams were built, they were abandoned—the yields from the valleys were too small, the hillsides had little spruce—and so, too, was the forested land abandoned. On almost all of the tract, lumbermen forfeited ownership by failing to pay taxes during the 1870s. Subsequently no one wanted to buy it from the state. The last interior tracts returned to the state in the 1890s. The region's hardwoods were never logged, the surrounding valleys were only sparsely settled, and the interior valleys were deserted by all but a few fishermen.

This forested legacy characterizes this 105,000-acre wilderness at the end of the twentieth century. As the logging camps were abandoned, so were the logging roads, except those used by a very few sportsmen. The dearth of roads accounts for the lack of trails. Only the old road to Whitehouse penetrates the wilderness and only one major trail crosses the area—the twenty-two-mile section of the Northville-Placid Trail, which follows a series of abandoned logging roads and touches a half-dozen lakes and ponds. The rest of the old roads are almost invisible. Rising from lowlands near the Sacandaga River nearly 2500 feet to the top of Hamilton Mountain, most of the interior is a high plateau, punctuated by little-known mountains (only one, Cathead, has a marked trail).

The West River Road, which once ended at the sportsman's lodge known as Whitehouse, gives access to wonderful forests and river views. As much as the lack of trails does to shelter the area, a few additional deep woods trails would not affect the integrity of the area. More trails would let more people enjoy the beauties of Groff and Abner Creeks, the draw between Wallace and Three Ponds Mountains, or the mysteries of King Vly. A through route from the North Branch of Stony Creek to Devorse Creek would surpass the Northville-Placid trail in wildness.

Views and Visits

The best views of this wilderness are along N.Y. 10 between Arietta and Piseco. Here the West Branch mirrors tall pines that permit only glimpses of Loomis, North Branch, and Three Sisters Mountains. Paddlers find wildlife, colorful meadows, and quiet as they navigate the dozen miles of slow-twisting meanders.

The Benson Road reveals the broad sweep of Three Ponds and Wallace mountains. Cathead, in the southeast, is a fire-tower mountain with views across the forested expanse. The drive up West River Road from Wells is generally too far from the river to enjoy all but a few glimpses of it, but it leads to a parking area at Whitehouse. Walk south along the Northville-Placid Trail and cross the West Branch on the wonderful, shaky bridge to marvel at the way the spruce and hemlock forests have recovered in 120 years. They are so impressive it is hard to believe they were ever logged. Even a short stroll on the trail to the north leads to some of the great hardwood stands that typify this wilderness.

The Silver Lake Wilderness has only one trail, so this wonderful waterfall on Groff Creek, like all the other falls in the area, has no access route. It is a bushwhacker's delight.

Photograph courtesy of Wayne Virkler.

6 Shaker Mountain Wild Forest

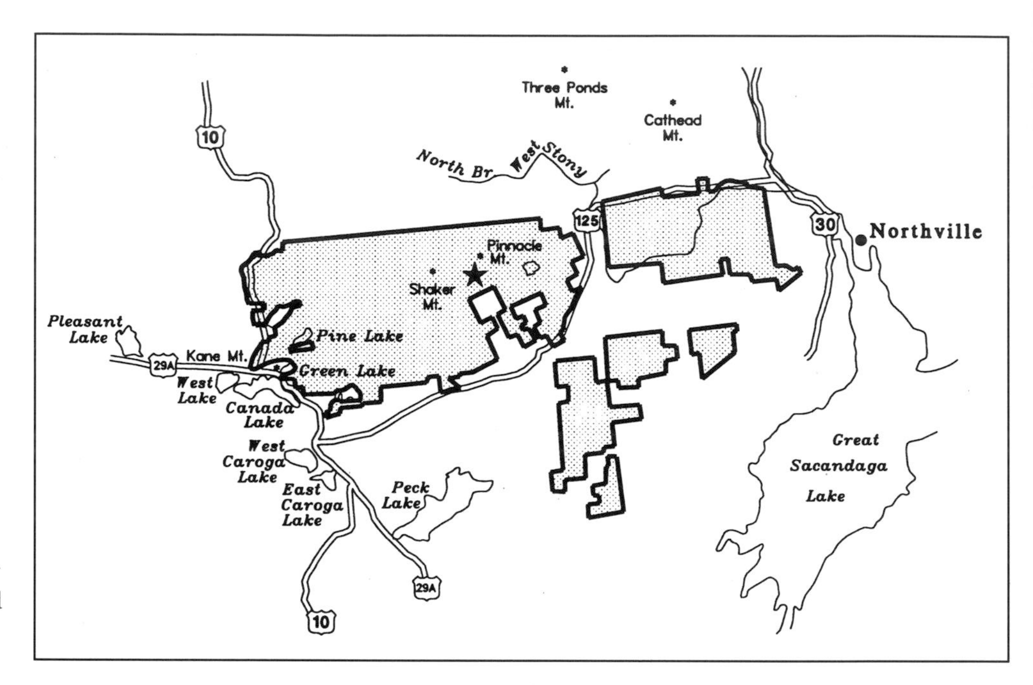

Some parcels of land are precious parts of our quilt because like a favorite dress they are old and worn, yet treasured. Shaker Mountain Wild Forest is like that—its land is old and worn, yet treasured, and what a history it has! Lying on the very southern edge of the Adirondacks, this land in Fulton County was not included in the original Blue Line that defined the Park. As a result, many parcels that had been obtained by the state for nonpayment of taxes were offered for sale again. These lots in Chases's Patent and parts of Glen, Bleecker, and Lansing Patent were the sources of hemlocks for a tannery, hardwoods for a chair-rail factory, and timber for shingles and sawlogs. Old mill sites lie at the end of Pinnacle Road, at Holmes Lake, on Holmes Lake Outlet, and at Frie Flow. Many of these mills date from the end of the nineteenth century, when state land in Fulton County was resold.

Many lots were logged a second time. The second-time-around loggers ignored a few lots whose pockets of noble forest contained hemlocks and hardwoods that graded into lowland ashes and oaks. Most of the other lots were sold again to the state, but choice pieces that have been held as Forest Preserve since the 1870s stand out as gems in a little-known wild forest.

Few wild forests are harder to discover than Shaker Mountain. Old roads and private tracts break up the tract, its core is totally surrounded by private lands, and all its secrets are well kept.

Stretching north of Gloversville and rising sharply in a series of small hills west of the Sacandaga Valley, the forests beckoned early settlers. Farmers in Benson cleared fields for sheep, and sometime long ago, probably in the first decades of the nineteenth century, the slopes north of Stony Creek burned. The pioneering popples—big-tooth aspens—are well into old age and are among the largest specimens of that species in the Adirondacks. This tired land, where sheep once grazed, was all abandoned before 1877.

Old roads to the farm sites and factories are still visible; some are snowmobile trails, some just places to find to walk in the woods. And, if you can find them, a few lead to the lots that were never resold. Scattered south of Stony Creek, and on higher slopes on both sides of Pinnacle Road reaching up to the mountain range that forms the northern border of Fulton County, lie forests of amazing stature. This choice 550-acre tract south of Stony Creek encompasses a large, complex, unnamed mountain. Secreted in valleys along the northern border are stands of hemlock and pine and yellow birch and maple that are as tall and straight as in any Adirondack place I have ever seen. The hemlocks on this patch, which has

Stony Creek traverses the Shaker Mountain Wild Forest. Much state land borders the creek, but that land in turn is surrounded by private parcels. Only a few small tracts of state land provide access to the many stretches of the creek that once harbored settlements.

Photograph courtesy of the author.

been protected since 1877, may have been stripped of their bark more than 120 years ago, but nothing else seems to have been touched.

The lack of trails is notable—snowmobile trails do not touch the wilderness keeps secreted within this wild forest.

Views and Visits

Walk north along Pinnacle Road from the parking lot for Chase Lake. Otherwise you have to bushwhack to find the most spectacular lots.

7 Ferris Lake Wild Forest

Ferris Lake is my favorite wild forest, probably because it was the first forest I knew. Even as a child I walked the short paths to its numerous ponds. Later I skied the longer trails that followed old logging roads. I have enjoyed the surfaces packed by snowmobiles, just as loggers had enjoyed the way these roads were iced for winter travel with horse-drawn sleds of logs.

I bushwhacked to the few cliff-tops among the relatively low mountains. Every summer for years I took my children to the waterfall on the East Canada Creek, near the confluence with Brayhouse Creek. Now they take their children for picnics and frolics under the waterfalls, slides down the small chutes, and "baths" in the potholes above the falls.

Memories are not all that lure me back to the place. Years ago I was impressed with the magnificent forest that surrounds the mid-northern section of the Powley-Piseco Road, the road that bisects the wild forest and connects Stratford with N.Y. 10 south of Piseco. In my first guidebook, I wrote that for those who could not walk to distant places, a drive along this road would substitute for a wilderness experience. How little I knew then of the reasons why this trip seemed so wonderful.

In the last few years I began to study forests somewhat systematically, researching the length of time they had been held by the state and discovering what had been logged and when. This

Ferris Lake Wild Forest has old-growth stands of hardwoods that vibrate with color in fall. Patches have been state land for so long that even the shores of Clockmill Pond, which once had a mill, are gloriously wooded. Photograph courtesy of the author.

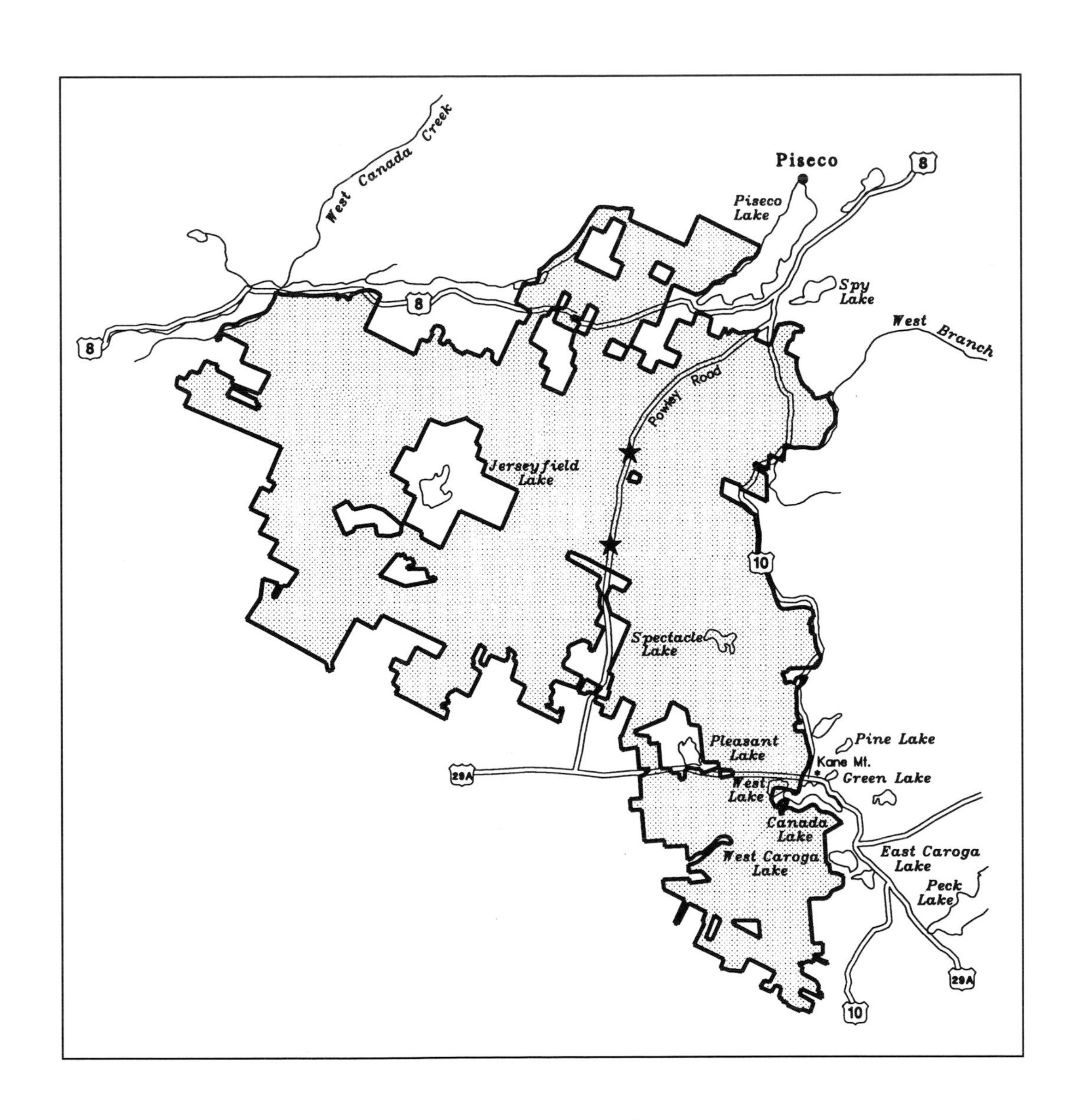
West Canada Creek
Piseco
8
Piseco Lake
Spy Lake
West Branch
8
8
Powley Road
Jerseyfield Lake
10
Spectacle Lake
Pleasant Lake
Pine Lake
Kane Mt.
Green Lake
29A
West Lake
Canada Lake
East Caroga Lake
West Caroga Lake
Peck Lake
29A
10

work led me to understand that most southern Adirondack forests logged before the middle of this century had been logged for softwoods only. Where only hardwoods grow, on the deep, rich lower slopes, almost no logging ever occurred. Before 1890, when loggers began to take smaller trees for pulp, only the largest spruce and hemlocks were ever cut. At first I thought that the only unharvested spruce stands would be found on tracts in northwestern wilderness areas with forests that had been sold to the state as virgin.

Imagine my surprise and delight when I discovered that a huge swath of the forest along the Powley-Piseco Road had never been cut and that it had the most spectacular spruce stand in the Adirondacks. No wonder I had thought the place was special! Gradually I learned that these stands extended miles to the west of the road and in places a fair distance to the east. How exciting it was to take botanists to these stands and have them exclaim that they were more beautiful than any spruce stands in the Five Ponds Wilderness!

Such an old-growth spruce stand has tall, straight, but not very large trees—spruce never get very large. In fact, a thirty-four-inch diameter tree on the side of the road is within a few inches of the diameter of the largest spruce ever cut in the Adirondacks. The forest floor is deep with mosses, covered with sorrel and woodferns. The place is damp and dark and mysterious, with the skeletons of fallen trees outnumbering those still standing.

If those old forests were all you might find in the Ferris Lake area, it would be considered extraordinarily special, but there is more. Among the numerous lakes and ponds are Blind Man's Vly, Christian Lake, Clockmill Pond, and Red Louse Lake. Besides the Brayhouse and East Canada Creeks are the headwaters of Black Creek, Fourmile Creek, and Goldmine Stream with its lovely waterfall. Long, relatively level trails follow old roads and connect most of the ponds. These trails invite the skier in winter but discourage the hiker in summer unless a summer drought dries up the mud that fills the ruts and hollows of these old roadways. Best of all is a drive through the forest on a day when it is too wet for hiking anywhere else. The road that keeps this area from being designated a wilderness takes you right into a wilderness the planners overlooked.

Popular destinations accessible from N.Y. 10 or N.Y. 29A include Nine Corner, Broomstick, Good Luck, and Jockeybush Lakes. Infrequently visited lakes like Spectacle, Dry, Dexter, or Third and Fourth, have longer trails that are ideal for cross-country skiing.

Views and Visits

Picnic at the Potholers, a couple of hundred yards upstream on the East Canada Creek near the confluence with Brayhouse Brook. Drive the Powley-Piseco Road and stop in the spruce forests a mile or more north of the iron bridge over the East Canada Creek at Powley Place.

Ferris Lake Wild Forest also has virgin patches of spruce with giants like this one gracing its old-growth stands.

Photograph courtesy of Chuck Bennett.

8 Black River Wild Forest

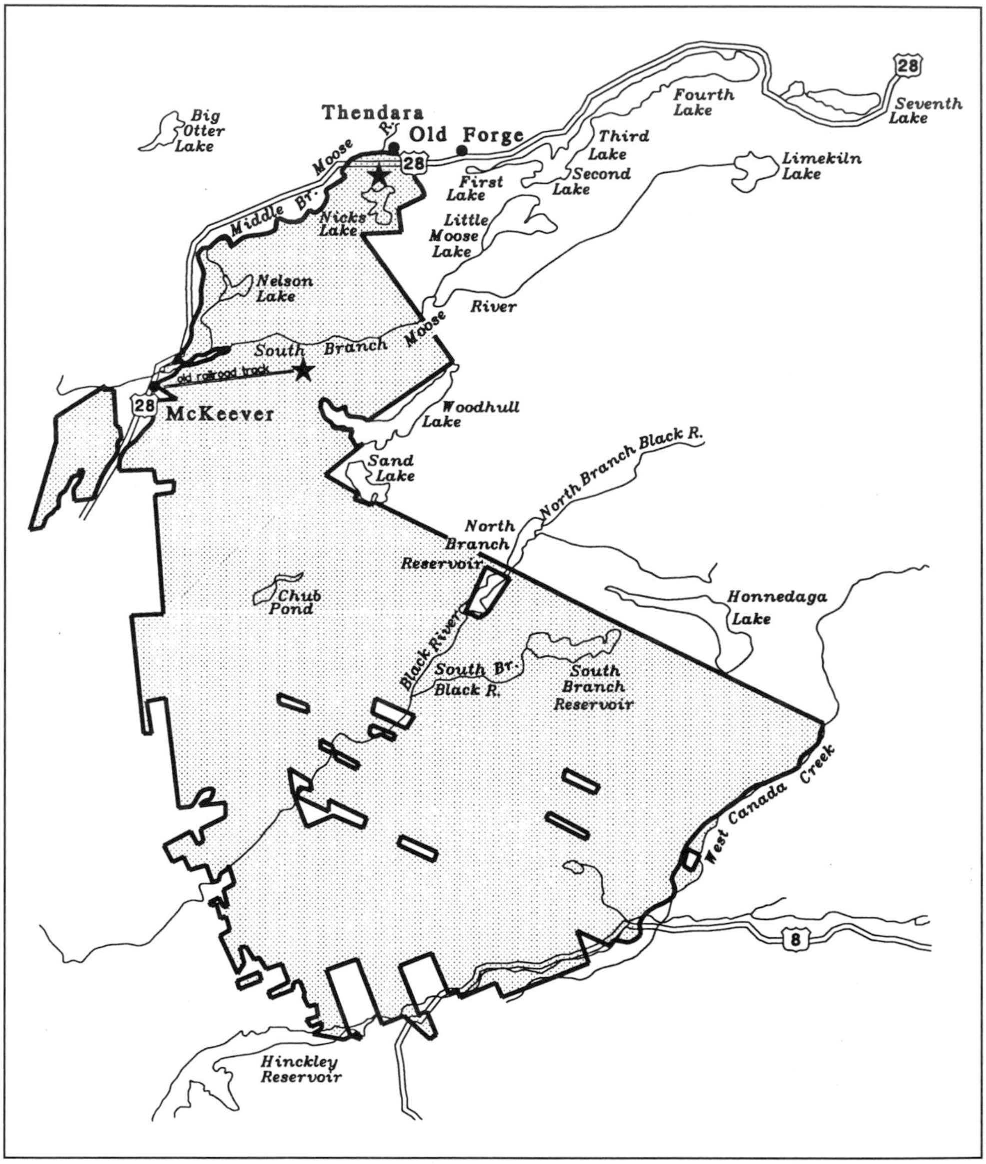

This huge—121,000 acre—wild forest is split by North Lake Road and peppered with small private inholdings but in every other way it is a wilderness, not just because of its size, but also because of its origins. Perhaps the relative flatness of its terrain—it has only one significant mountain, Woodhull—made planners think "wild forest" and "snowmobiles." Certainly the presence of roads giving access to private lands in scattered inholdings detracts from its status as a wilderness. With Ferris Lake and Wilcox Lake Wild Forests, Black River Wild Forest is part of the triumvirate of great southern Adirondack tracts. And, like its sister wild forests, Black River is itself made up of patches that would be considered wilderness anywhere else.

Rivers define this region: it straddles the Black River and encompasses parts of the Moose River and tributaries of the West Canada Creek, which borders the wild forest on the east.

Many of its lakes were dammed as a way of controlling floods along the lower Black River and providing power to its downstream mills in summer droughts. The dam on North Branch Reservoir was completed in 1856, followed by a dam on South Lake in 1859 and on Woodhull Lake in 1859. Most of the surrounding land was logged—some for softwoods, and with the railroad extension from McKeever, some for hardwoods as well. The forest yielded logs for pulp, railroad ties, and hemlock bark. The railroad grade is

A beaver meadow and water cascading over a tiny beaver dam provide a restful opening in an otherwise densely wooded area. Photograph courtesy of Lee M. Brenning.

Nature changes the landscape in dramatic ways as demonstrated by the blowdown from a tornado that traced a path across the Black River Wild Forest. Photograph courtesy of Lee M. Brenning.

now a hikers' highway that leads to trails to Remsen Falls on the Moose and to Woodhull Mountain.

Many logging roads were reopened in the 1930s by the Civilian Conservation Corps (CCC) for fire protection. Roads once led to many of its lakes; today they are trails to Chub and Gull and Sand Lakes. So a cursory observation would seem to confirm its designation as a wild forest.

Still, the heart of the Nobleboro Patent, south of South Lake and far from the West Canada or Black Creek, was abandoned for taxes early on. It is a wild place of old hardwood forests and extensive wetlands. Bordered on the north also by the private lands around Honnedaga Lake, it contains beautiful Jocks Falls on the outlet of that lake. Those falls are accessible through public land only by the most adventurous hikers and fishermen.

What the logger did not take in this southwestern corner of the Park, nature did. The blowdown of 1950 was particularly severe. The big wind of August 1972 devestated the Woodgate area. A tornado-like storm in July 1984 ripped a swath through the wild forest, clearing a 350-foot wide swath from Chub Pond toward Little Woodhull Lake and beyond.

Views and Visits

Walk the trail along the abandoned railroad grade from McKeever to the Nelson Falls trail or sample trails that surround Nicks Lake Campground.

9 West Canada Lakes Wilderness

This sprawling, hourglass-shaped wilderness is the second largest in the Adirondack Park, encompassing 157,000 acres. The Northville-Placid Trail crosses it north to south and a circle of trails loops around the spectacular lakes that fill the waist of the hourglass. The central core with its ring of trails has long had the reputation as a sportsman's paradise and as a far-away but reachable wilderness. Brooktrout, Whitney, Pillsbury, Mud, South, Big West, and the Cedar Lakes were settings for fishermen's tales.

The wilderness has few entrance trails. The Northville-Placid Trail approaches the wilderness from the south and Piseco or from the north and the Cedar River Flow. A trail leads to its interior from the western edge of the Moose River Plains, another leads from near Lewey Lake over the eastern mountain range that otherwise is all but inaccessible. The easiest entrance is at the eastern portal of the wilderness, along an old trail from Speculator that reaches Sled Harbor through lands

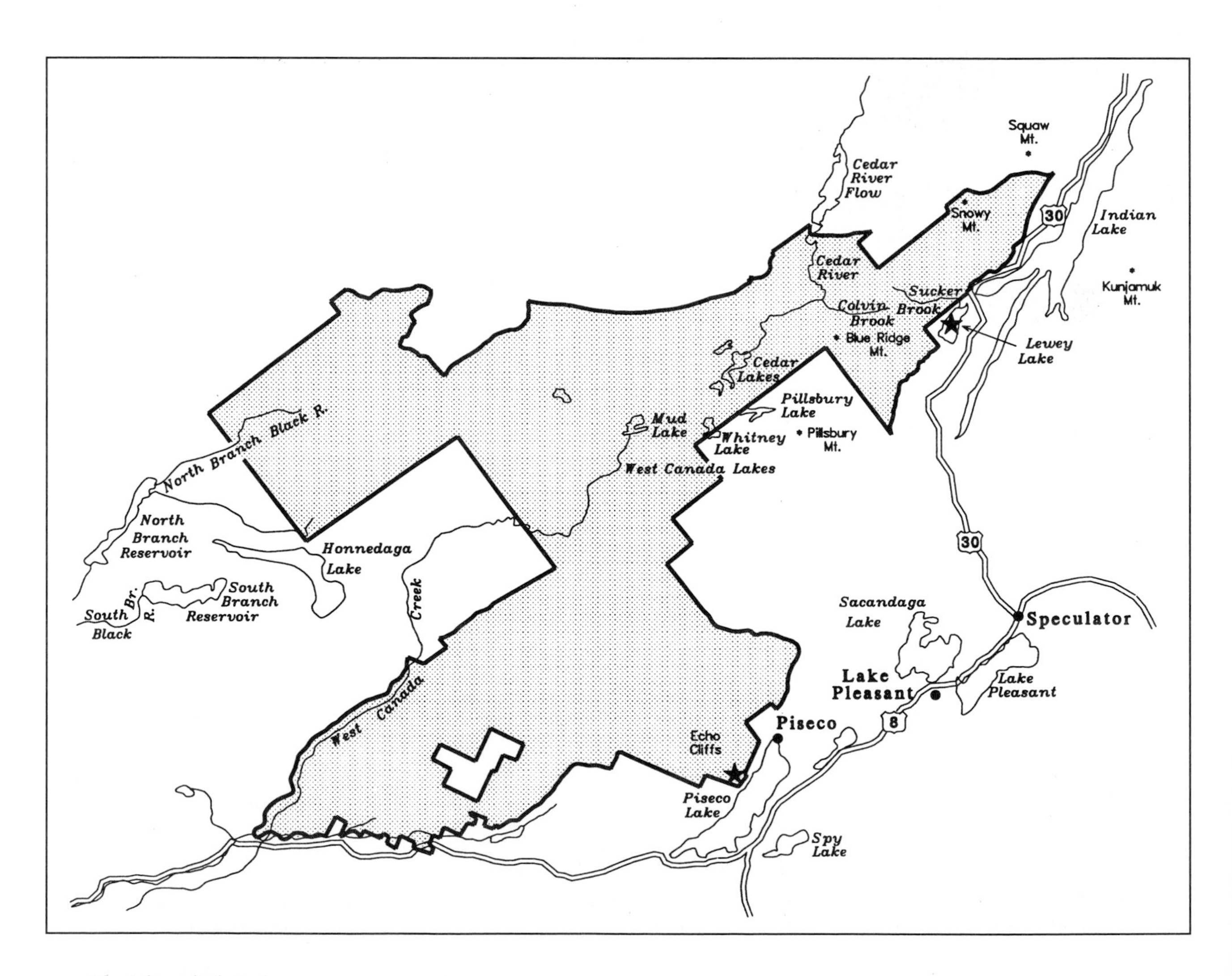

Snowy Mountain overlooks Indian Lake and dominates the range of mountains that guards the eastern flanks of the West Canada Lakes Wilderness. Photograph © 1998 by Carl E. Heilmann II.

owned by International Paper Company (IP). One other trail through those lands leads directly to the Northville-Placid Trail east of Spruce Lake.

The history of the protected interior of this wilderness is fairly recent but much fabled. Loggers came up the West Canada Creek and down the Cedar River, but for the most part, they had no easy way to harvest the vast forests where softwood stands were scattered on ridges and in marshy valleys between hardwood slopes. Even the Jessup River, which could carry logs to the Indian, never floated logs from the slopes of Lewey Mountain.

A part of the old Albany Road to Sacketts Harbor, a bare track cut out after 1812, traversed the region from Speculator north toward the North Branch of the Moose River. That road all but disappeared within the next twenty years. French Louie, a famous hermit and sportsman, was said to have been the first to cut a trail from Sled Harbor to the Cedar Lakes in 1873. His route probably followed the same path as the old Albany Road and the fact that he is credited with cutting the trail attests to the overgrown state of the old Albany Road. Before 1873 the interior was virtually untouched except by the rare hunter or fisherman.

On the headwaters of the West Canada Creek, any logging that occurred before 1870 was done south of the central core of lakes. A square north of the Cedar Lakes was logged between 1906 and 1920. The roads left by these logging operations invited twentieth-century fishermen, who were drawn to the cabin French Louie built on the north shore of Big West Lake. As the cluster of lakes became an ever-greater lure for fishermen, drawing them in increasing numbers, early float planes replaced the horses and wagons that bumped along the deteriorating logging roads. By that time, the Conservation Department was manning an interior ranger's cabin at the site of Louie's cabin.

Almost all of those who visit this wilderness use the trails that lead to the center ring. Vast forests to the north and south of that ring probably receive fewer visitors than they did a century ago. The Little Great Range (Pillsbury, Cellar, Blue Ridge, Lewey, Snowy, and Squaw Mountains) that flanks the eastern border of the wilderness is cut through in only two places—the trail past Sled Harbor and the trail through the pass between Cellar and Lewey Mountains along Sucker and Colvin Brooks. That trail is perhaps the wildest marked route in the Adirondacks.

In the south, rimming Piseco Lake on the northwest, lies another range of mountains. Echo Cliffs on Panther Mountain in that range provides a surprisingly wonderful view after a fairly short climb. Behind this range, T-Lake and the notably tall T-Lake Falls are remote destinations. The falls, the tallest in the Adirondacks, plunges 350 feet over a dangerous, rounded escarpment.

There never have been any substantial fires anywhere in the West Canada region. Vast portions of the interior have never been logged, or have been logged so lightly and so long ago for spruce alone that almost all the forest but the area around the Cedar Lakes and to the north can be considered old growth. The one exception is the tract bordering IP's land, where a land exchange in the 1980s gave IP virgin lots and the state acquired cut-over lands of equal value. The resulting better-defined boundary will only improve the protection afforded this wilderness and its forests, which are gems in the patchwork of the Forest Preserve.

Views and Visits

The West Canada Lakes Wilderness is so well protected by mountains and the great distances required to reach the interior that no easy ways to enjoy its secrets exist. From a canoe on Lewey Lake or the Cedar River Flow you can gaze up at the eastern and western flanks of the Little Great Range. Pillsbury Mountain has limited views without its fire tower. Echo Cliffs has a wonderful view, mostly of the adjoining Silver Lake Wilderness.

The Cedar River passes below the Carry Lean-to and provides water access to the northern border of the West Canada Lakes Wilderness.

Photograph courtesy of Lee M. Brenning.

10 Jessup River Wild Forest

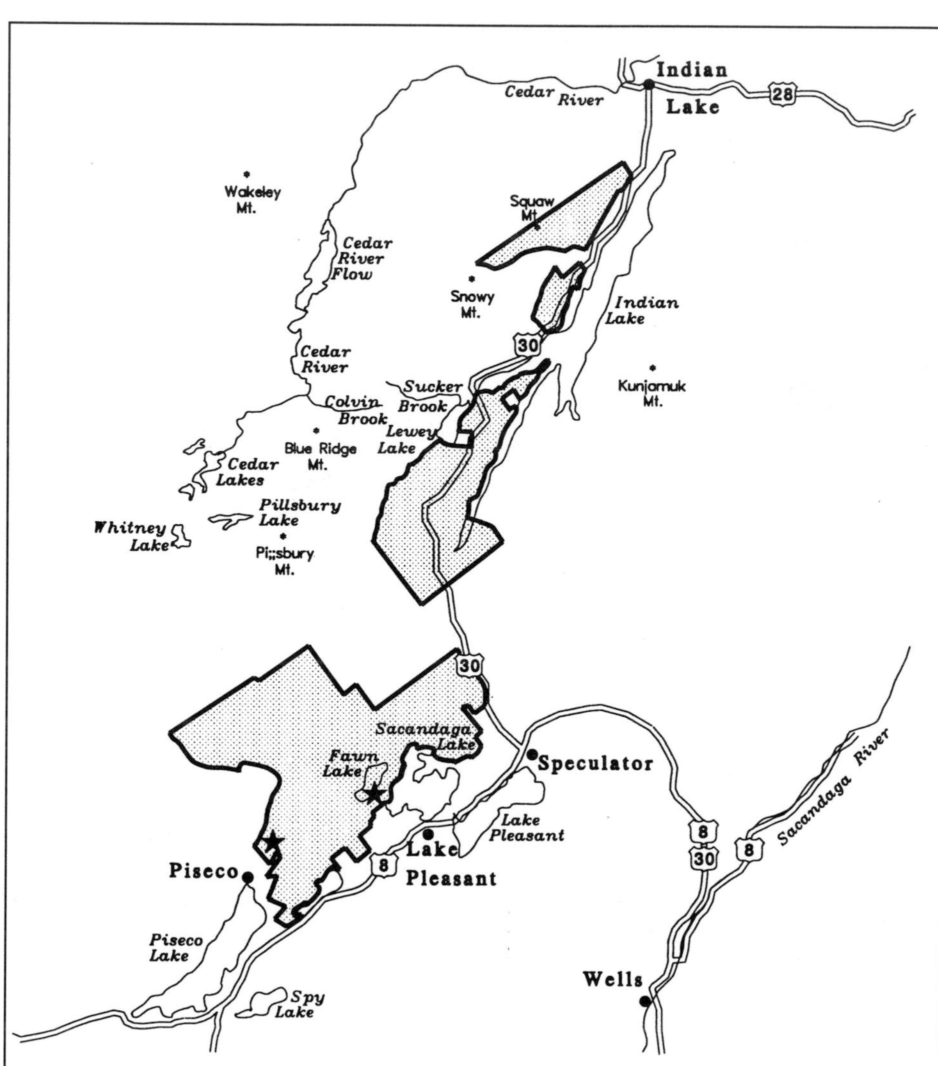

Tiny Jessup River Wild Forest stretches in patches north from Piseco to Indian Lake, chinking in spaces between wilderness areas and private lands. Originally it contained a checkerboard of lots interspersed with squares held by International Paper Company. The checkerboard pattern was erased when IP swapped land with the state in order to consolidate its holdings. The tract still contains a few virgin lots that were not a part of the swap as well as cut-over lands south of Sled Harbor. But it also has some very important forests: most of the Sukey and Newbold tracts south of Lewey Lake straddling N.Y. 30 and the Jessup have been owned by the state since 1877, and almost all of that has never been logged. The forest here differs very little from those in the wilderness area to the west that stretches along slopes from Cellar and Lewey Mountains toward Snowy Mountain.

North of Piseco and Oxbow Lakes, lots in Moose River Township 9 were logged in the last third of the nineteenth century. According to descriptions of the time, these lots were purchased by the state, and the logging was very light. Majestic forests now stretch all the way from Piseco Airport around to Fawn Lake.

Dominated by wetlands and spruce and hemlock forests surrounding Falls Stream, Burnt Place Brook, and the lower Jessup River, the only reason not to call this a wilderness is its small and fragmented nature.

Views and Visits

Walk along the ski trail north of Piseco Airport or canoe Fall Stream for a flavor of the region.

Mist, calm water, and images of rocks paint a quiet landscape on a soft fall morning at Lewey Lake in the Jessup River Wild Forest.

Photograph courtesy of Chuck Bennett.

11 Moose River Plains Recreation Area

The premier recreation area for hunters and those who like to camp from their cars is an area that presents two faces to the public. The public face, the roadside, the much frequented ponds, and the heavily logged forests, are all most people see. That face conceals a hidden one of deep woods, trailless mountains, cliffs, and bogs that few see. Most visitors never stray far from the major access roads that traverse the Plains and link it to highways, and to many they seem like corridors in the wilderness road. Beyond their borders lie many destinations that are as remote and wild as you will find in any wilderness.

Moose and Manbury Mountains are tall enough to sprout caps of spruce and balsam as dense as any high peak. The cliffs behind Mitchell Ponds or the even taller ones that rise above the Moose River on Mitchell Ponds Mountain itself are hidden view points known only to a few hardy souls. It is true that a great deal of the Plains was logged and parts were probably burned to produce browse for deer. Certainly the Plains were overrun by deer, which kept the forest from recovering. But even the drive along the roads reveals towering pines and tamaracks that are the culmination of pioneering stands. This is wilderness in the making.

The western part of the Plains, with two wonderful lakes, Squaw and Indian, has a long way to go before the forest there recovers. It was logged right up until the middle of the twentieth century by Gould Paper Company, which was among the first to use heavy equipment to build roads and haul logs to its mills. Thickets of wrist-sized balsam

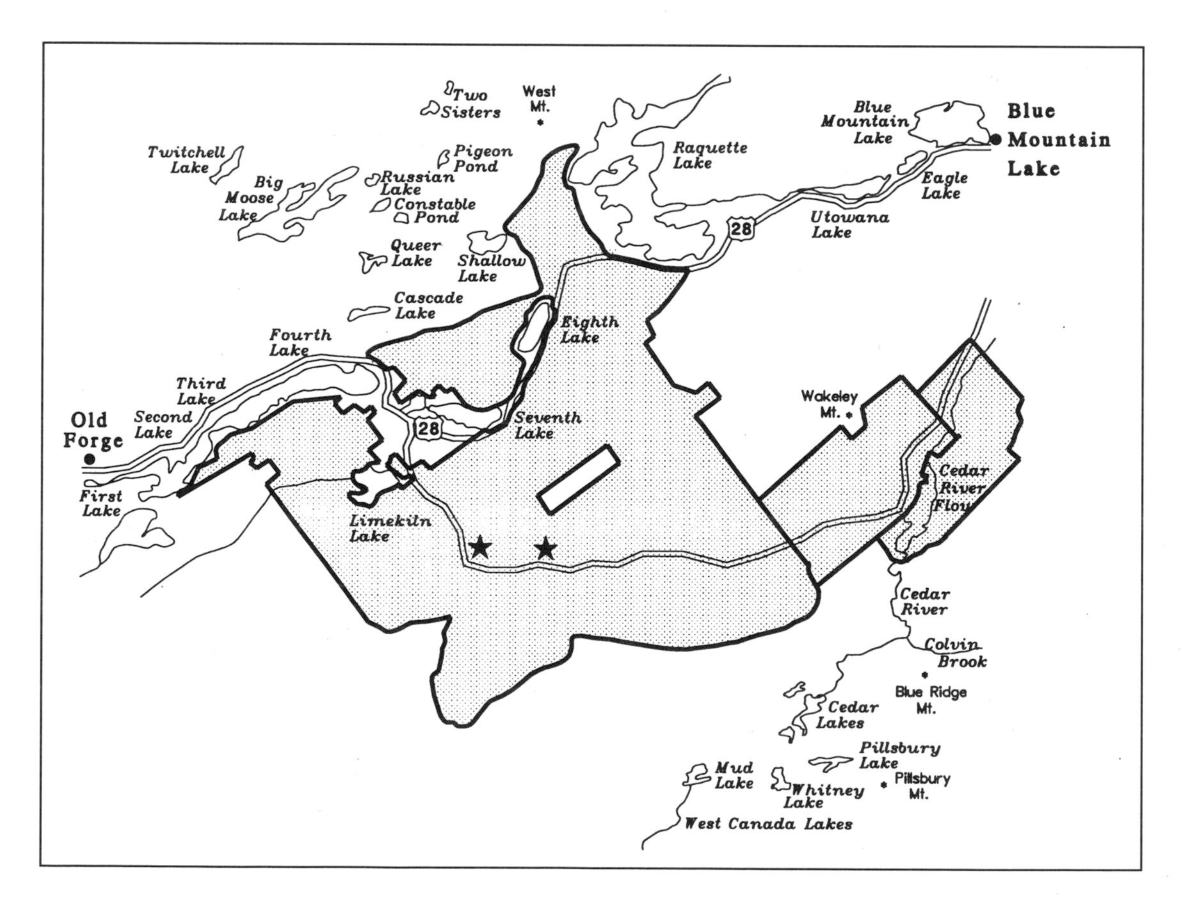

A pair of ponds stretch out below Mitchell Ponds Cliffs. This distant view does not reveal the wonderful plant life that surrounds the ponds that are just two of the myriad beauties in the Moose River Plains Recreation Area. Photograph courtesy of the author.

and cut-over hardwoods will take a long time to grow.

Marshes and wetlands, some even close to the road, reveal a botanist's delight in their diversity of ferns and wild flowers. These pockets prove once again the ability of the landscape to recover.

Views and Visits

A drive through the Plains Road from Wakely Dam on the Cedar River to Inlet, with a short walk to Helldiver, Lost, or Icehouse Ponds samples the beauty of the Plains. A longer but still very easy walk to Mitchell Ponds reveals many different trees and ferns.

12 Ha-de-ron-dah Wilderness

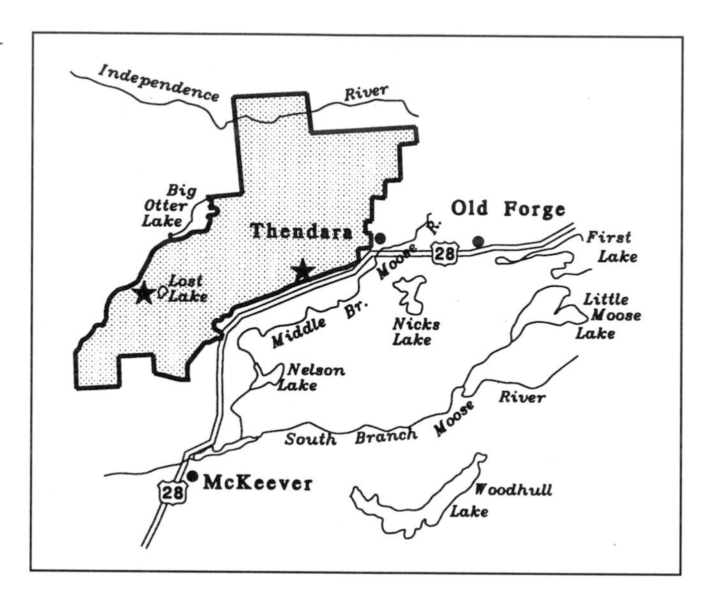

This small, less than 27,000-acre wilderness straddles the Herkimer-Lewis county line. It is part of the 210,000-acre tract acquired by John Brown in 1798. From the time it was first surveyed into townships just before 1800 until it became state land, this tract has had a tragic history. First, the tract failed to support the families attracted to the settlement by John Brown's son Francis. The dam at First Lake and the saw and grist mills powered by the lake's water soon fell to disrepair. A son-in-law, Charles Frederick Herreshoff, tried to revive the community and arranged to have a road built from Moose River Settlement to the tract. Brown's Tract Road survives as a trail through the wilderness, but all else is gone, including Herreshoff's manor house and the forge he built to process iron from a mine that turned out to have inferior ore. Herreshoff's suicide after the mine flooded ended the community for all but guides and visiting sportsmen.

This wilderness was logged at least once, and the southern three-quarters was burned in a 1903 fire that swept over 25,000 acres. This fire, started by the railroad that lies just outside the southeastern border of the wilderness, was one of the largest single fires ever to devastate the Adirondacks. It burned all the way from the railroad corridor north to the Otter Pond Road and in places crossed that road.

The state acquired most the tract shortly after the fire: nearly 12,000 acres in a tax sale that resulted from a default by Taggart Paper Co.; 4,600 acres purchased in 1909 in Township 1; and nearly 6,000 acres in Township 7 from Lyon deCamp, whose grandfather, Lyman Lyon, had acquired large forest tracts in the western Adirondacks for his mills and tannery on the Black River. On his death, each of his three daughters received parts of his land. His daughter Julia, who married W. S. deCamp, inherited a large tract north and west of Old Forge, part of which is now in the Ha-de-ron-dah Wilderness.

The fires that turned these tracts into land no one wanted was a strange beginning for a future wilderness. The fires are evident in stands of uniform-sized, second-growth trees; in huge meadows of hay-scented ferns that probably preclude the regeneration of denser forests; and in the preponderance of such pioneering species as cherry and birch and poplar.

The roads survive in a fairly dense network of trails that connect almost every one of its twenty lakes and ponds. These trails take you on long but gentle walks through forests that inspire with nature's restorative powers.

Views and Visits

A parking lot 3 miles south of Thendara on N.Y. 28 is opposite the principal trailhead for the region. A short walk north along the entrance trail takes you to an intersection with the trail that follows John Brown's Road.

A rustic bridge takes hikers along the Lost Lake Trail in the Ha-de-ron-dah Wilderness. Photograph courtesy of Dale W. Richards.

13 Independence River Wild Forest

It is easy to understand why the Independence region did not receive a wilderness designation. The dry, sandy soils on the east support scrub growth and open fields instead of great forest stands. Numerous roads in the western part have become snowmobile and horse trails. Private lands interrupt the wild forest, but their access roads lead to trailheads for state land.

The wild forest sweeps gently up from lowlands not far from the Black River and rises north and east to low hills covered with dense forests and interspersed with wetlands. Low, gentle hills outline sluggish streams that converge to a series of west-flowing rivers, each with lovely waterfalls. The Independence River with Gleasman's Falls, Otter Creek with Shingle Mills Falls, Crooked Creek with its falls—each river has a special stretch approached by hiking trails.

In the move east from scrubby fields and sand barrens to tall, but young, woods, the transition to wilderness is blurred. Long trails that follow recently abandoned logging roads between ponds and rivers are best suited for snowmobile travel.

Number Four Road defines the northern border of the wild forest—it offers a wild drive and one of the few places where you can be certain of seeing ravens in the tall pines. Midway on the road between Stillwater Reservoir and Crystal Dale lies Francis Lake, an outstandingly beautiful body of water edged with pine-covered eskers.

Gleasmans Falls is just one of many waterfalls in the Independence River Wild Forest.

Photograph courtesy of the author.

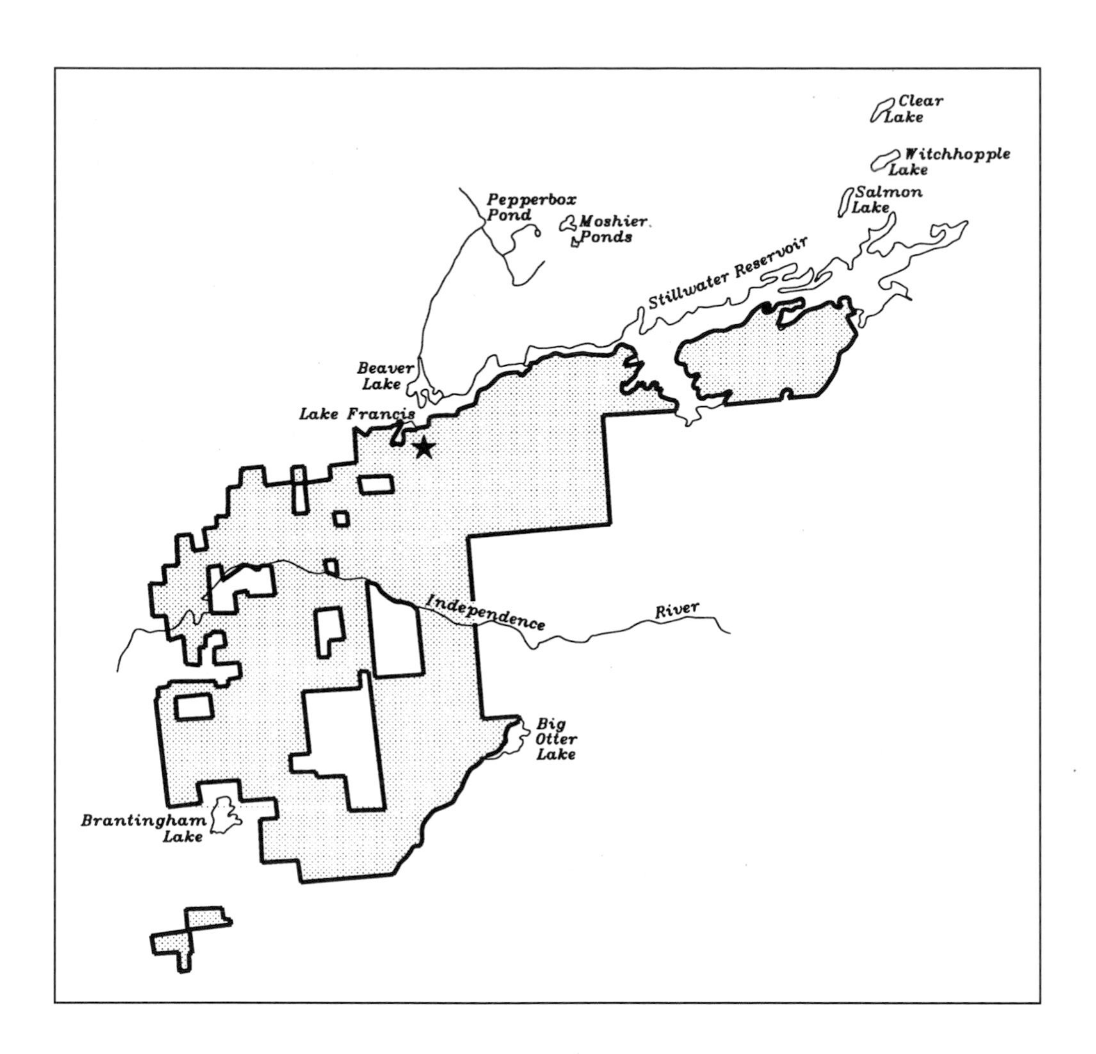

Numerous ponds, old roads, and snowmobile trails lead south from that road. Regardless of the private inholdings that are concentrated in the south and east of the forest, the Independence River Wild Forest has large patches of Forest Preserve and is one of the wildest regions in the park.

Views and Visits

Drive Number Four Road and launch a canoe on Francis Lake.

14 Fulton Chain Wild Forest

If this small wild forest were a piece of fabric, you might describe it as moth-eaten. The holes are all the private lands surrounding such lakes as Rondaxe, Twitchell, Big Moose, and the western end of Fourth Lake. The Pigeon Lake Wilderness borders it on the north and east; private lands lie to the west; and the Moose River Plains adjoins it on the south. The public land is stretched over a series of sinuous ridges of hills with sharp cliff faces looking south.

In the midst of one of the Adirondacks' busiest highways, this stretch north of N.Y. 28 has more Forest Preserve trails per square mile than almost any other place. There are bicycle routes along the old railroad bed that led from Thendara to Raquette Lake as well as mountain trails and ridge walks. Snowmobile trails lead out to private lands, the railroad, and Big Moose Lake.

Pockets of old-growth forests lie between hillsides burned long ago and once again covered with tall trees. Part of the wild forest lies in tracts sold to the state before they were ever logged. Perhaps some of the easiest trails to view magnificent forest stands lie within its boundaries.

Two beautiful trails through magnificent forest take you to the outlet of Sis Lake for this view toward Bubb Lake. Photograph courtesy of Lee Brenning.

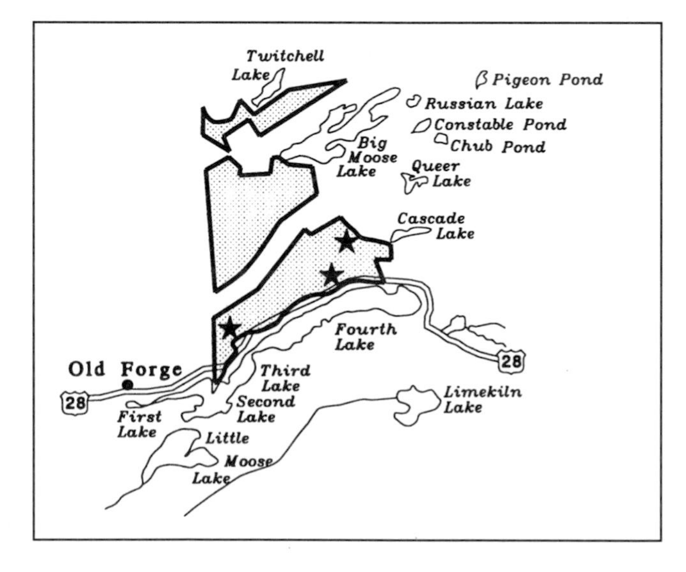

Views and Visits

Drive the Uncas Road, which borders part of the wild forest on the north. Climb Bald Mountain or any one of the little peaks north of the Fulton Chain. Walk around Moss Lake or to Bubb and Sis Lakes to enjoy old-growth forests.

Some think the views from Bald Mountain are among the best in the Adirondacks. A brilliant fall day and colorful leaves enhance the view over First Lake. Photograph © 1998 by Carl E. Heilman II.

15 Pigeon Lake Wilderness

The Pigeon Lake Wilderness is unique in that its 50,000 acres contain a greater proportion of virgin lands than any other wilderness area. Its eastern two-thirds lies in Totten and Crossfield Townships 40 and 41, lands sold by William Seward Webb to the state in 1897 and 1898 as virgin lands. Their "virginity" was analyzed along with their potential for producing merchantable spruce in 1901 and 1902 by the Forest Commission under a grant from the federal government. Uncut stands of spruce brought Webb an extraordinary profit for the time; the $6 to $6.50 an acre for lands acquired by payment of back taxes set a record for the time.

The wilderness reaches east to private lands around Raquette Lake. These lands were owned briefly before 1850 by Farrand Benedict and his relatives. Their failure to pay taxes on the land was key to their preservation in a pristine state. Only a narrow band around Raquette Lake was logged.

Long, east-northeast trending valleys are filled with marshes and wetlands and towering evergreens. The land would be valuable for its forests alone, but its beauty is enhanced by numerous lakes—Shallow, Pigeon, Cascade, Queer, Chub, Russian, Upper and Lower Sister, Terror—and ponds such as East and Constable.

A wedge in the northwest corner of Township 41 that contained Lake Merriam was logged between 1904 and 1914. The only fire in the area was a small one on the slopes south of the Sister Lakes.

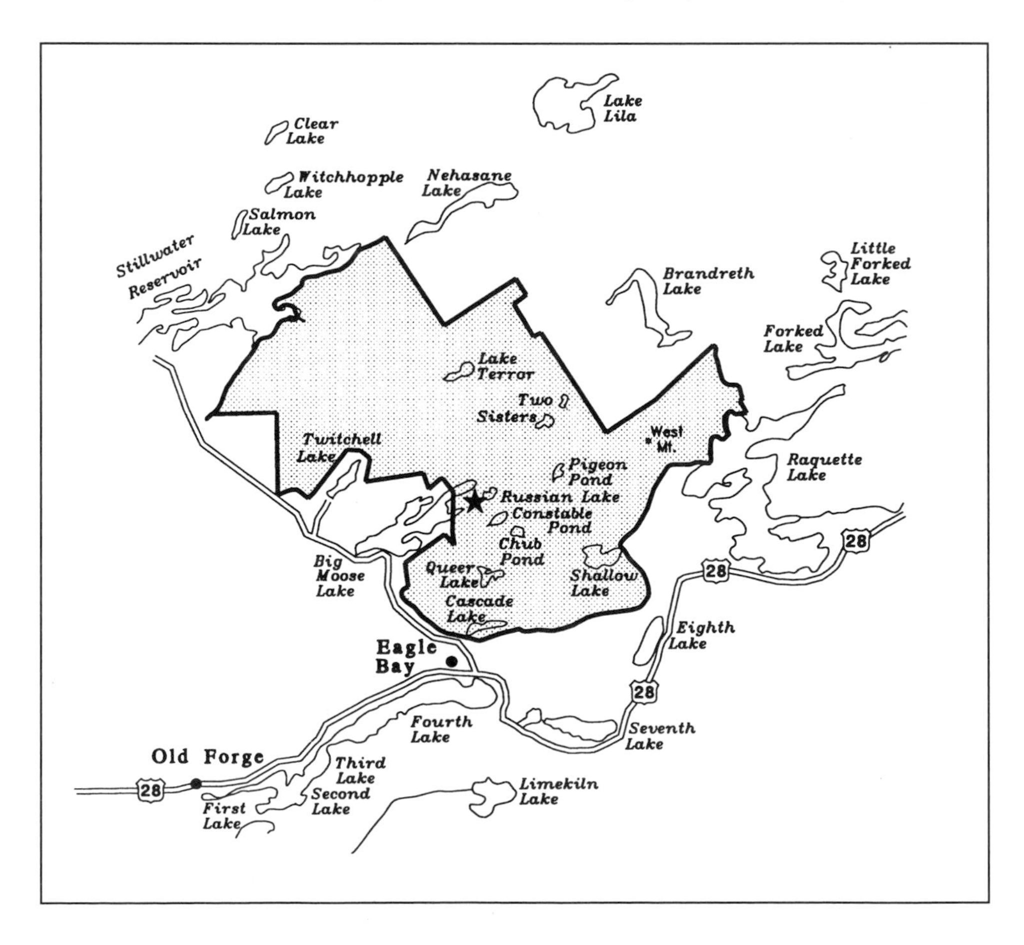

A small part of the wilderness south of the railroad and Beaver River was logged and became part of the Forest Preserve at a later date. But what man has left undisturbed, nature has not. Many parts of this magnificent forest were knocked down in the 1950 blowdown, although today you can see little damage. The 1995 storm heavily damaged forests on the slopes of West Mountain, the eastern sentinel of the wilderness.

There are forests here that are more spectacular than any in the Adirondacks. The trail to Russian Lake is a gem of pines, spruce, and hemlocks. The trail from Windfall to Queer Lake has stands of yellow birch of extraordinary girth. And although the region does not have a great many miles of trails, it has a network that takes you to almost all its lakes and ponds as it samples the varied forest stands that make up this truly natural wilderness.

Views and Visits

All approaches are long but one rather complicated trip is certainly worth the effort. If you can arrange a boat or canoe trip across Big Moose Lake to its southeastern inlet, there is a dock at the beginning of a mile-long trail that leads to Russian Lake. It offers the best sample of great forest in the park.

Lonesome pine at Queer Lake echoes the stillness of winter. Photograph courtesy of Dale W. Richards.

Meadows, wetlands, and marshes characterize this wilderness, which started out as the only designated trailless area in the Adirondack Park. The Wilderness Lakes tract north of Stillwater Reservoir, which separated the original Pepperbox from the Five Ponds Wilderness Area, was recently acquired by the state. This tract was split between the two areas, divided by the logging road through Shallow and Raven Lakes and north. The western half, added to Pepperbox, will be managed to maintain that area's intended trailless state, even though it will be years before the hand of man will fade.

Even without the trailless designation, natural forces would continue to dominate this wilderness. Shallow ponds, beaver flows, and meadows bordered by dense alders interrupt a landscape of knobby little hills. Parts of the terrain are indistinguishable from each other, except for the steep slopes that identify the south faces of those hills. Chains of ridges and hills border the western part of the Pepperbox, again with features blending together to make travel by map and compass difficult.

The ponds are devoid of fish, the forests are only now beginning to recover from logging, and there are enough insects to discourage people while sustaining the area's rich bird life. This will be a wilderness where nature truly reigns.

A walk in deep woods highlights unmarked paths in wilderness areas such as the Pepperbox.

Photograph courtesy of the author.

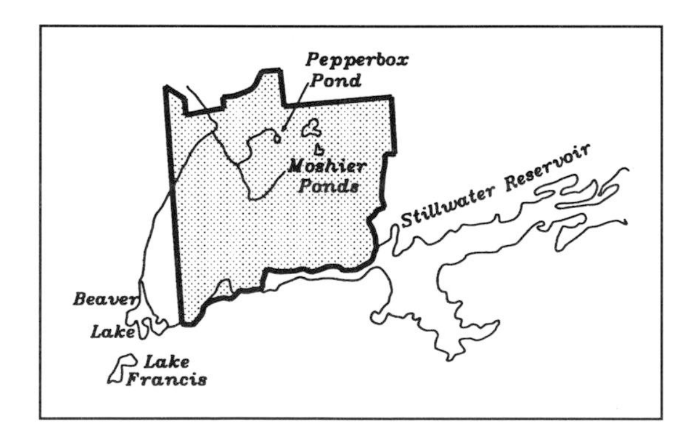

17 Five Ponds Wilderness

The Five Ponds Wilderness Area, which covers over 95,000 acres, started out with a truly mixed parentage that conjures images of both rape and virginity. Virginity because so much of the southern part of the region has never been logged; rape because the railroad-assisted logging in the north portion stripped vast stands of pines from the area south and west of Cranberry Lake in an incredibly short period—seven years at most. The southern boundary of St. Lawrence County cuts through the wilderness and separates the two regions with these vastly different histories.

The largest single purchase to create this wilderness was a 40,000-acre tract acquired in 1896 as virgin land from William Seward Webb. This broad dagger reaches all the way down to Stillwater Reservoir and encompasses most of Totten and Crossfield Township 43, Township 51 or the Triangle north of

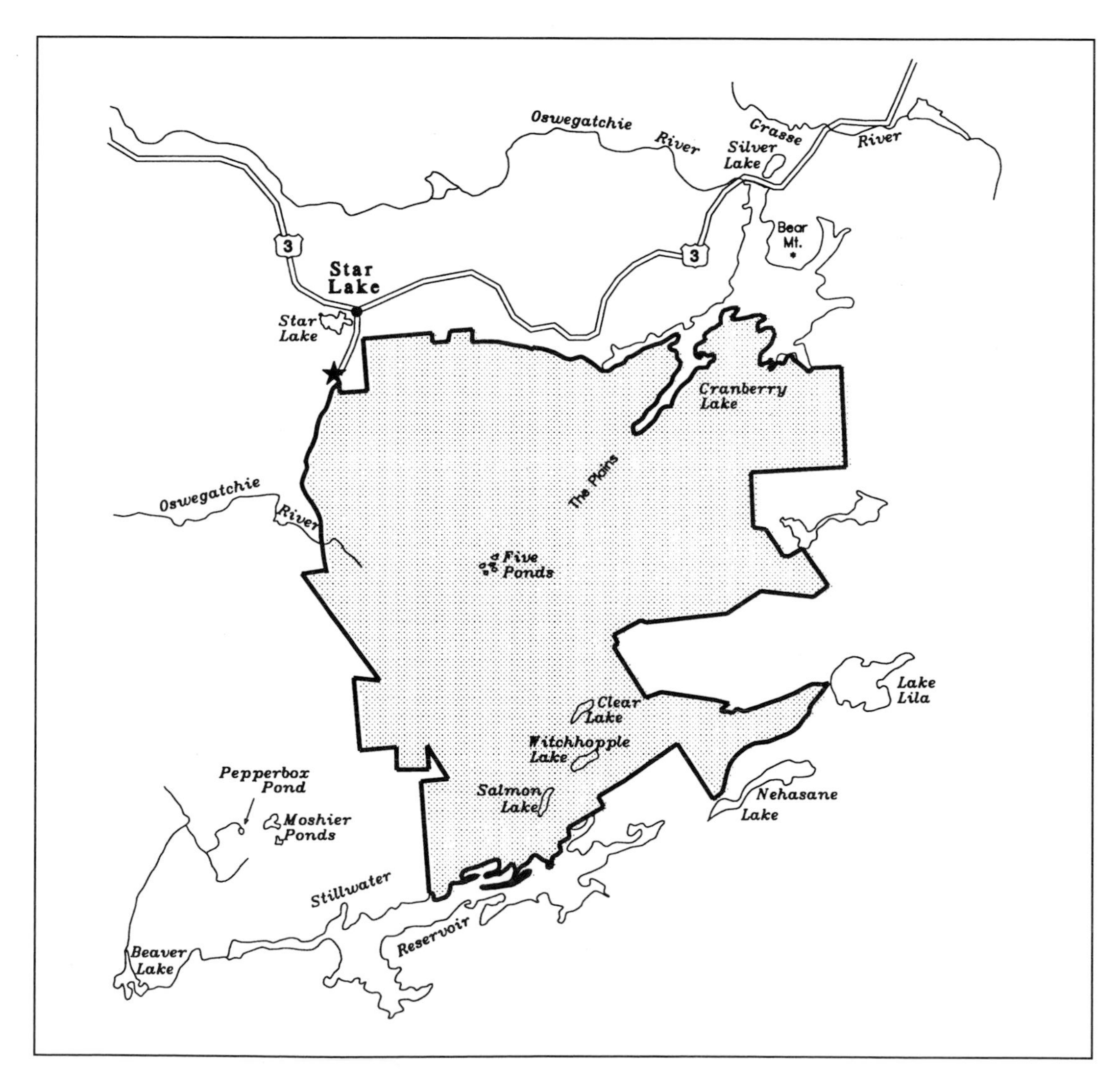

Township 38, and part of Township 42 in Herkimer County. Added to this were the tax-sale lots in St. Lawrence County just to the west of Cranberry Lake, so that over half of the future wilderness had become state land by the turn of the century.

In the next three decades, cut-over tracts were added, Emporium and Rich lumber companies selling the largest. Only 10,000 acres were added in the next six decades, to bring the wilderness to its current size.

Forests cover all but two regions—the Plains, which has been a natural, open, grassy area, and a much smaller natural plain north of Salmon Lake.

The region was all virgin early in the twentieth century when railroads were built to haul out the forest riches. Rich Lumber Company extended a railroad along the Oswegatchie River; Post Henderson Company had a railroad farther west; and Emporium Lumber Company's railroad approached from the east. Fires followed Rich Lumber Company's logging in the northwest corner of the future wilderness—in 1912, 1913, and 1915. The wilderness continued to grow as loggers and fires laid waste to the forests in the north, rendering them worthless to the lumbermen and ripe for state acquisition.

Roads and railroads in the St. Lawrence County portion became the region's trails. The first Adirondack Mountain Club trail, the Robinson Trail, was proposed as a north-south traverse of the region; it has long since faded into wilderness. All that survives is the trail from Stillwater north past Salmon Lake to Witchhobble Lake and on to Clear Lake; it is the only marked trail in the southern part of the region.

Sand Lake is one of the most remote places in the Adirondacks, remote because it lies near the heart of the Five Ponds Wilderness. Photograph © 1998 by Carl. E. Heilman II.

The wilderness grew because William Seward Webb felt that the 1893 dam on the Beaver that created Stillwater Reservoir had made it impossible to log his lands. He claimed he could not use his railroad for this tract, but he certainly could have used the river. No matter, he decided to sell.

The eastern half of the Wilderness Lakes Tract that borders Stillwater Reservoir was appended to the Five Ponds Wilderness, even though in the 1980s, roads were pushed through the tract for one last timber harvest.

In ways, the Five Ponds Wilderness is a crazy quilt itself—soft patches with the velvet plush of natural grasslands; striped sections with parallel eskers forming ridges like corduroy, where each ridge is covered in the velvet of deep forest; prickly pieces rough with pines and evergreens; soft pieces gently topped with leafy hardwoods.

In the summer of 1995, windstorms savaged this wilderness, leaving blowdowns so severe that many parts are impenetrable and will remain so for decades. This event added another texture to the Five Ponds Quilt—a deep tweed in which a dense, dark brown layer of stumps, trunks, and branches is flecked in pale green of new sprouts and trees.

The Oswegatachie River circles through the Five Ponds Wilderness, then twists and turns through broad wetlands below High Rock. This stretch is a favorite canoe route.

Photograph courtesy of the author.

This description does not do justice to the forested ridges and eskers and beautiful ponds that dot the area. Nature's fury has enveloped the five tiny ponds in the heart of the area with a savage wildness that will be a magnet for adventurers for years to come.

Views and Visits

A short trip to see the forces of nature starts along the snowmobile trail on the western border of the wilderness, leading south from Star Lake. Even a short walk along it will give you a sense of nature's destructive power.

18 Aldrich Pond Wild Forest

Aldrich Pond Wild Forest is the northwestern buffer for the Five Ponds Wilderness. The state acquired the 16,000 acres that form the core of the area from the International Paper Company in 1986. Recent purchases and easement additions of Lassiter Land in Watson's East Triangle have extended this area into a wild forest of significance. The southern border of the region reaches all the way to the wilderness tracts that lie to the north of Stillwater Reservoir. A network of roads, albeit rough, four-wheel-drive tracks at best, make even the most remote corners accessible to outdoors people.

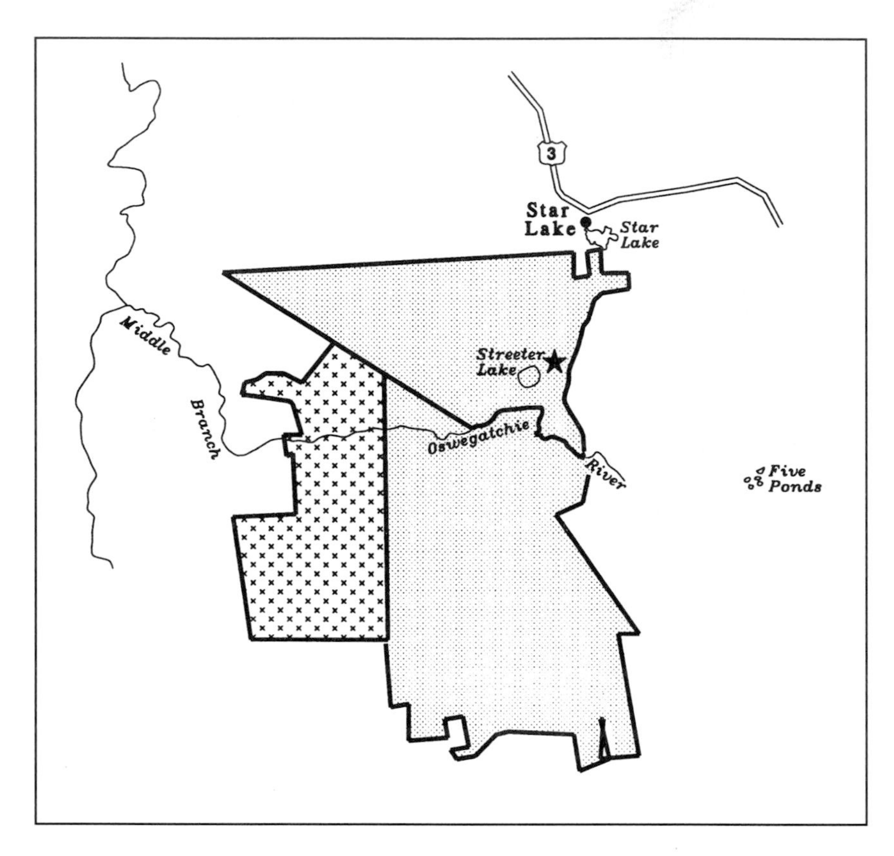

The Middle Branch of the Oswegatchie, with its myriad tributary streams, winds across the region, leaving the wild forest in the rush of Rainbow Falls, a spectacular drop in a narrow gorge. Streeter Lake and the ponds to the northwest of it flow north into the Main Branch of the Oswegatchie. The region's amazing variety includes long, flat chains of beaver meadows, ridges of eskers, old logging roads, recovering farm fields, heavily logged stands of forest, and deep woods, including even a remote piece of old growth.

Today, remains of logging dams abound. Old settlements like Jayville with its nearby iron mine have faded back to forest. Railroad beds that once served logging companies are now snowmobile trails. Hunters enjoy telling of the jackworks (staging and loading areas for logs) and the hermits' and hunters' hideaways that occupied the abandoned logging camps. Names and adventure tales relieve the unmodulated terrain.

The state lands are fast becoming wilderness, but a very accessible wilderness. The private lands, including easements, mean that vehicular access will continue. The Long Pond–Belfort Road is a rock-strewn adventure that swings through the wild forest. The tract's classification and these connecting roads will make this the most accessible wilderness in the Park.

Views and Visits
Drive through the northern part of the wild forest to Streeter Lake.

Recent state additions to the Aldrich Pond Wild Forest give access to spectacular Rainbow Falls on the Middle Branch of the Oswegatchie.

Photograph courtesy of the author.

19 Cranberry Lake Wild Forest

Cranberry Lake Wild Forest is another of the buffer regions separating private lands from wilderness, its pieces bridging the gaps between the two. This wild forest is overwhelmed by the Five Ponds Wilderness to its south. Dominated by the large, man-made lake with its state campground, the region has many trails—even a new boardwalk nature trail.

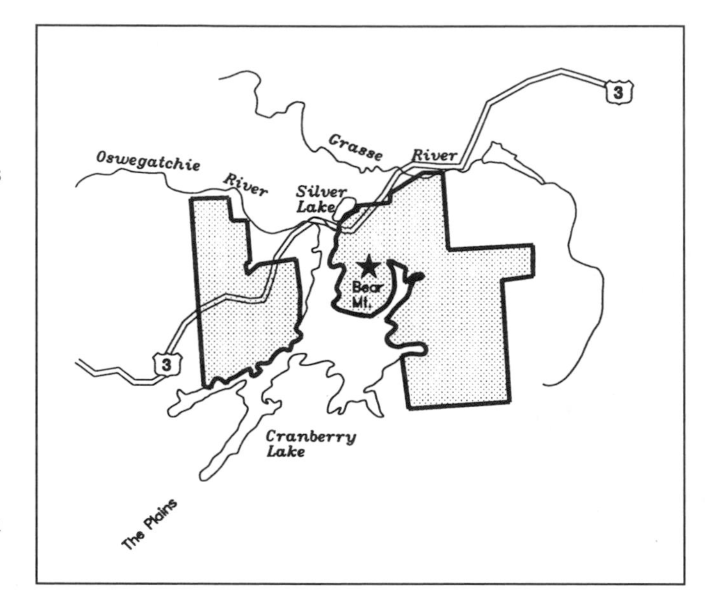

The trails are remnants of logging days, which did not begin in this area until the twentieth century. In 1913, the Emporium Forestry Company built the Grasse River Railroad leading west from Conifer to Cranberry Lake. This railroad enabled the company to harvest hardwoods as well as softwoods, and very quickly the 15,000 acres that comprise the core of the wild forest were stripped of timber and, by 1934, sold to the state.

Cranberry Lake was created by a dam on the Oswegatchie River. The first was built in 1867; a second, higher dam, was built in 1916 to supply power for downstream hydro plants. Motorboats ply the sprawling impoundment, reaching to trailheads on state land as well as the private cottages around the lake.

Boating, fishing, big game hunting, camping, and snowmobiling are the main attractions here.

Views and Visits
Climb Bear Mountain for a view across Cranberry Lake or walk the Nature Trail from the campground.

A beaver pond mirrors a wetland near Cranberry Lake. Photograph courtesy of Lee M. Brenning.

20 Horseshoe Lake Wild Forest

Horseshoe Lake Wild Forest is a forest of promise and access to wilderness. Lying east of the New York Central Railroad, the lands closest to the railroad suffered some of the worst of the 1908 fires. New acquisitions to the west of the railroad and along the Bog River and Hitchins Pond suffered even more. This tract was acquired by the state in 1985 and gives canoe access to Lows Lake. This small tract is designated a primitive area.

The Bog River was a highway for logs floated to mills at Tupper Lake and Piercefield. Today, Lows Upper Dam, built in 1907 by A. A. Low, holds back a nearly ten-mile flow that reaches all the way to Grass Pond at the edge of the Five Ponds Wilderness Area. Lows Lake, formed from the Bog River, is now a highway for canoe campers, who carry across the divide to the Oswegatchie River.

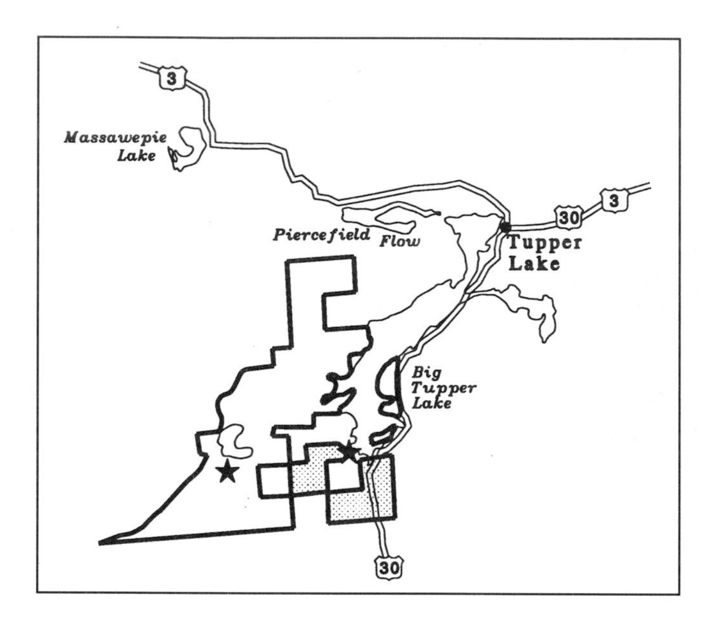

The county road (421) heading west from N.Y. 30 bisects the wild forest and circles south of Horseshoe Lake with its numerous campsites. It leads to trails to Big Trout Pond and Bridge Brook Pond. A great stretch of the wild forest is accessible by water only, across Tupper Lake. The sentinel of the northern apex of the triangle that forms the wild forest is Arab Mountain with its still-standing fire tower.

The wild forest is also bounded by two large private tracts—Whitney and Litchfield parks and several other working forest tracts. There are those who wish the private lands could become state-owned so the Horseshoe tract could be added to the Five Ponds Wilderness to the west, and someday that may happen. But with all the recreational opportunities at present, the status quo truly serves the public.

Rather than establishing one huge wilderness, the practical realities of developing this kind of reserve with many different kinds of access, including roads, makes sense. This has been done on the Horseshoe Lake Wild Forest and the private lands immediately to the west of it. Pursuing easements that limit development and recreational access to the lakes and mountaintops on surrounding private tracts to the south and northwest appears to be the best way of assuring future public values. It would also allow the private lands to continue in much needed forest production.

Views and Visits

Drive the road to Horseshoe Lake. Stop near the beginning of the road to look at the waterfalls just above where the Cold River empties into Raquette Lake. Drive on to look at Horseshoe Lake. Walk or drive to the lower dam on the Cold River and cross it to walk to Big Trout Pond.

The ledges above Lows Lake provide a grand vista across Hitchins Pond to the Horsehoe Lake Wild Forest, Horseshoe Lake itself, and beyond to Mounts Morris and Seward.

Photograph courtesy of Wayne Virkler.

21 Lake Lila Primitive Area and Little Tupper Lake

Beginning in 1890, Dr. William Seward Webb, a son-in-law of William H. Vanderbilt, acquired Adirondack tracts, which soon totaled 115,000 acres. A railroad man, he was impressed with the possibilities of a line that would connect Herkimer on the south with Malone in the north, opening up vast untouched forest reserves. He could not induce the New York Central to build the railroad, so he decided to build it himself. Because the state would not grant him a right-of-way, he had to purchase all of the right-of-way. Within two years, the line was completed, including a spur to Saranac Lake. This marvel of bridges and causeways through marshes passed through Webb's key tract, 40,000-acre Nehasane Park, which surrounded Lake Lila. Here Webb built a home and a private railroad station.

Influenced by Vanderbilt to consider scientific forestry on his tract, Webb arranged for Gifford Pinchot, a pioneer in scientific forestry, to survey Nehasane Park. Pinchot found a high proportion of merchantable spruce and devised a cutting plan to Webb's dictate that the logging would be done "without injuring the production power of the forest." Nehasane was logged for spruce in 1896–98. Fires, started by the railroad, burned on a third of Nehasane Park in 1903.

The forest was cut a second time, between 1913 and 1918, and hardwoods were harvested along with spruce. The railroad could carry the hardwoods to factories at Tupper Lake. The spruce harvest was considerably less than Pinchot had predicted—there was too much regrowth of hardwoods because the first cutting had been too severe. A third harvest in the 1930s left the forest with few softwood stands, and some of the tract was sold to the state. In 1978, the 14,600 acres surrounding Lake Lila was added to the Forest Preserve—the state bought the land for $100 an acre.

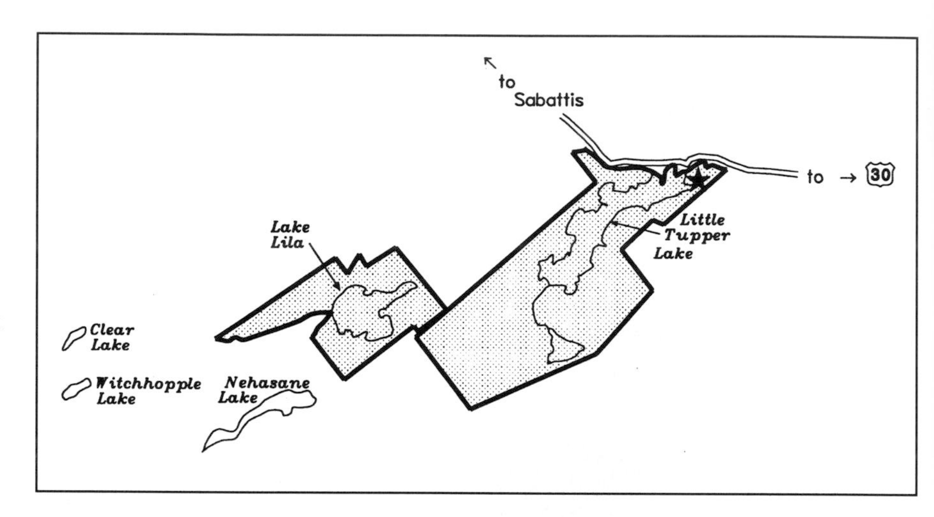

The tract remains a protowilderness because an access road leads to the remaining private lands to the northeast of the lake, but it is large enough to be a wilderness. The lake is beautiful, a gem; the forests are recovering, if slowly. But one wonders if this grand experiment in forestry will ever see a return of the softwoods for which it was famous. And, if they do return, how long will it take?

Touching the western corner of this primitive area is a 14,700-acre parcel that includes Little Tupper Lake. The state has just purchased this part of Whitney Park, adding a spectacular, watery gem to the Forest Preserve. A canoe route to Rock Pond with short carries leads to Shingle Shanty Brook and Lake Lila through some of the most scenic areas in the Park. This fifteen-mile route and 5.7-mile long Little Tupper Lake will be a paddler's dream. Easily accessible, the lake has tiny islands crowned with evergreens, deep bays, marshes, and a well-forested shore. It is quite easy to look beyond past logging and see the rest of this purchase becoming part of our forest treasure.

Whitney Park, assembled between 1896 and 1898 by William C. Whitney and lumberman Patrick C. Moynehan, encompassed mostly virgin tracts. The first logging ended in 1909, a second logging began in 1934, and both used horses to skid logs, thus protecting the

Rocks and grasses along the southeast lobe of Lake Lila invite exploring by canoe. Photograph courtesty of Wayne Virkler.

remaining stands. No trees were cut within two hundred feet of shore along the tract's many lakes and ponds. A third period of logging began in the 1950s and continued through the 1970s, but that time the logging was not as careful or restrained. The 1995 windstorm blew down much standing timber. No matter what nature and man has done, this forest, too, will recover.

Views and Visits

Drive west on County Route 10A from N.Y. 30 to Sabattis Road and the dam at the outlet of Little Tupper Lake or on to a canoe launch site.

22 Sargent Ponds Wild Forest

Three distinct patches have been stitched together to make the Sargent Ponds Wild Forest. The northern patch is a mountainous complex that rises to Owl Head Mountain. The southern part is a land of boreal swamps and forests of spruce and balsam, mixed with hardwoods. Bordered on the south by the Marion River and private lands, and a chain of lakes—Blue Mountain, Utowana, and Eagle Lakes. With a tiny railroad connecting the river to the lakes, this water route was once the principal approach to Blue Mountain Lake and its elegant hotels. The story of the hotels and the railroad is one of the major themes depicted at the Adirondack Museum at Blue Mountain Lake.

A string of hills rises north of the lakes and the river, then drops again to the Sargent Ponds and their marshy outlet stream. That stream empties into Raquette Lake, north of Tioga Point, the site of a nineteenth-century hotel. A smaller isolated patch of Forest Preserve is included in the wild forest. It encompasses the lower part of Brandreth Lake Outlet and land to the west of Forked Lake.

Good trails lead to the ponds themselves, sportsmen fly in to fish in them, and the swimming is excellent. What is needed is access to the southern approach to Lower Sargent Pond. The long trail to Tioga Point is little used—in fact, it is even hard to find where it crosses swamps and marshes.

Views and Visits

Walk the trail from North Point Road to Upper Sargent Pond.

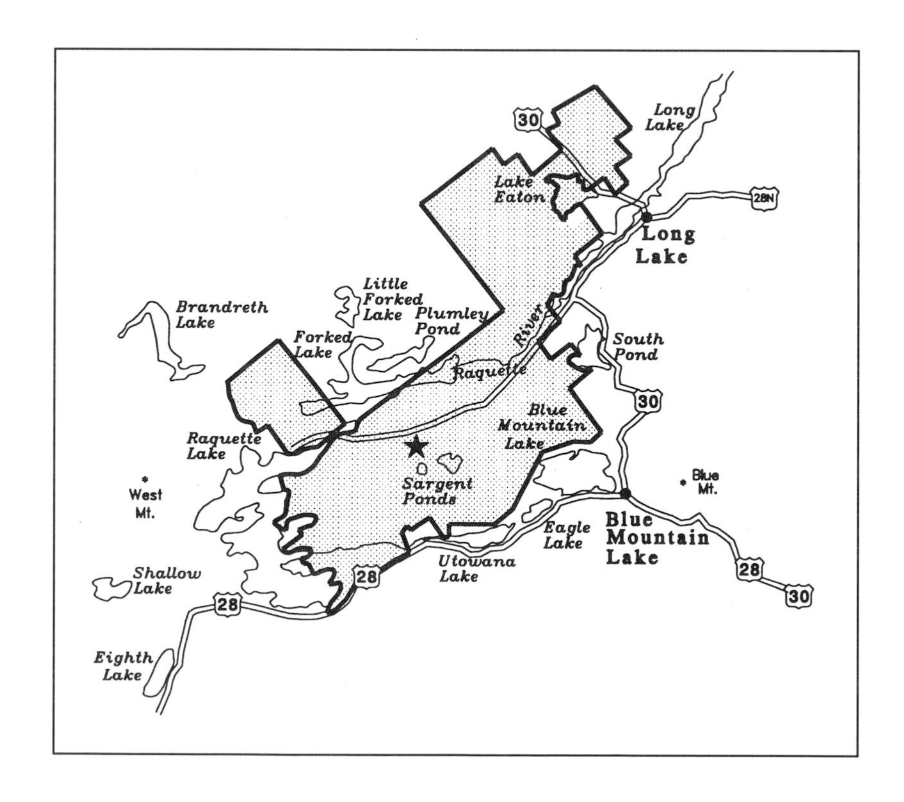

Cloud reflections at Lower Sargent Pond are undisturbed. The surfaces of the Sargent Ponds are often ruffled by swimmers.

Photograph courtesy of Wayne Virkler.

23 Blue Ridge Wilderness

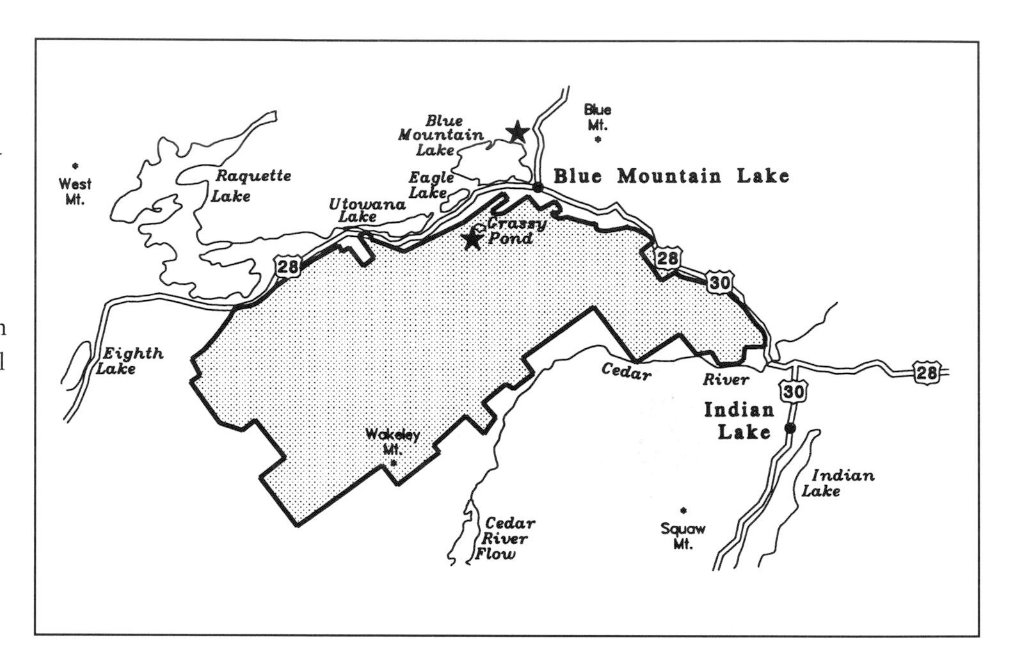

Blue Ridge is one of the most inscrutable wilderness areas in the Park. Old maps show no roads crossing its wooded passes except possibly the Albany Road, whose route is the subject of great conjecture. It is likely that this road was to the west of the Blue Ridge's western boundary. As there were no real maps when the route was first laid out, there was no way to define its course in relation to geographical features, which were largely unknown. Old-timers talk of another route from the Cedar River Road west, perhaps past Aluminum Pond, to Sagamore Lake, but all trace of that road has disappeared.

The western part of the wilderness lies in Totten and Crossfield Township 6, which, except for the tract around Sagamore Lake, was never logged. The eastern part lies in Township 34, where the portion south of Blue Mountain Lake and adjacent to the Rock River was logged by Jones Ordway in the late 1860s and 70s. That land and the southern slopes of the Blue Ridge almost down to the Cedar River Road were logged again in the early part of this century; logging continued for a few years after the state's purchase was completed in 1910.

In 1950, nature struck a particularly severe blow on the old forests in the western part of the wilderness, some of which were heavily laden with virgin spruce stands. The blowdown of that year decimated these stands, and the tangle from this destruction adds to the region's impenetrability.

A series of trails to small ponds, some linked to the Northville-Placid Trail, which crosses the northwestern corner of the wilderness, barely touches the forested core of the wilderness. The trail follows in part the road built by Ordway to connect his lodge near Wilson Pond with the Cedar River. The region has only two mountain trails, to Sawyer Mountain and to Wakely Tower, which is actually outside the wilderness.

The wilderness is bounded on the west by the road leading to the great camps built by William West Durant at Sagamore Lake and Lake Kora; on the north by N.Y. 28, where it parallels the Eckford Chain of Lakes and curves south toward Indian Lake; and on the east and south by private lands bordering the Cedar River and Moose River Plains roads.

It may be too much to hope for a trail that re-creates the old Albany Road, but it is exciting to contemplate one that reroutes the Northville-Placid Trail north from the roads near Cedar River Flow by circling west around the western flanks of Wakely Mountain, then northeast through the high valley between the Blue Ridge Range and Metcalf Mountain. That route leads directly into the existing trail well south of Stephens Pond. In addition, a loop

Cascade Pond nestles beneath the Blue Ridge Range in the wilderness of the same name. A lovely trail loops from Lake Durant to its shores.
Photograph courtesy of Chuck Bennett.

connects Stephens, Cascade, and Rock Ponds with the west end of man-made Lake Durant. Lake Durant, originally Marsh 34, was the staging area for logs shipped down the Rock River to the Hudson and the mills at Glens Falls with which Ordway was associated. The only other trail in the wilderness leads to Grassy and Wilson Ponds on the east slopes of the ridge, and parts of it also date to old logging roads.

Views and Visits

The broad sweep of the Blue Ridge itself is heavily wooded and will probably always remain trailless. It is visible across Lake Durant from the parking area on N.Y. 30 and from the grounds of the Adirondack Museum. The waters of Wilson Pond mirror a part of the Blue Ridge range, but a fairly long trail leads to that pond.

24 Blue Mountain Wild Forest

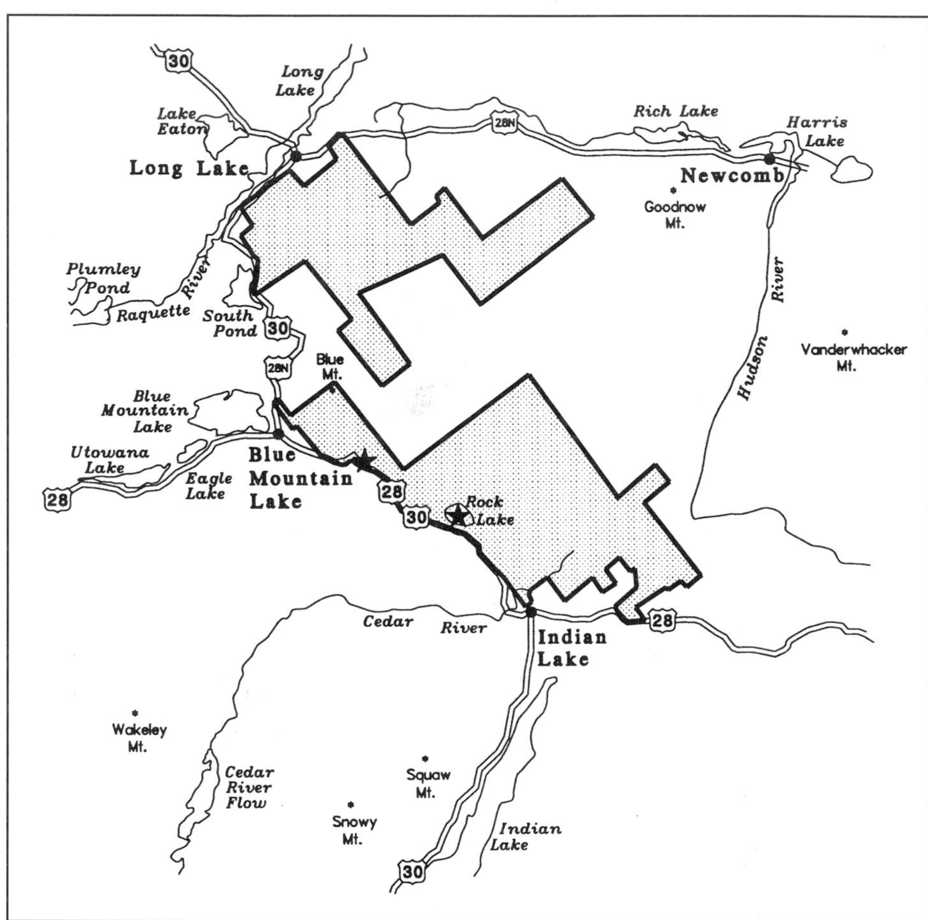

Private lands split this wild forest, depriving the segments of wilderness designation. The northern patch is a high realm of mountain tops, encompassing East Inlet Mountain on the west to the Fishing Brook Range on the east. The Northville-Placid Trail traverses a lofty pass in that range. State land surrounds Tirrell Pond, so that trail can continue along it, but the trail is routed on private lands both north and south of the pond. Only the summit of the eponymous Blue Mountain, not the beginning of its approach trail, lies on state land.

The southern patch is dominated by Rock Lake and the loop of the Rock River, which is making its way from Lake Durant to the Cedar River. That river once resounded with the logs harvested in Township 34 near Blue Mountain Lake. It is hard to imagine that water released from dammed-up Lake Durant could float logs down the river that more than lives up to its name. The Stark Hills and Ledge Mountain join other cliff-faced little hills in accenting the peculiar topography of the southern patch. All the strange slopes surrounding Unknown Pond were burned. Extending across the Cedar River, the Blue Mountain Wild Forest includes that river all the way to Elm Island. Mysterious McGinn Mountain with rampart cliffs surrounding its summit punctuates the wild forest's eastern boundary. That area, too, was burned.

Over 20,000 acres of cut-over land in Totten and Crossfield Township 17 were purchased by the state in 1900 for less than two dollars an acre. It was land that had lost all its value for lumbermen, but this was a remarkable low price compared to the six dollars an acre the state was paying at that time for other such tracts.

The trail to McGinn Mountain needs to be renewed, but the rest is so fragmented it is difficult to recommend ways of improving access to this wild forest.

Views from Blue Mountain attract hordes of visitors. This view is across Tirrell Pond to distant mountains. Photograph courtesy of Chuck Bennett.

Views and Visits

Blue Mountain looms over N.Y. 30 with its cliffs visible from many places. The northern trail to Rock Lake is a short walk, but the trail leads not to views of Blue Mountain, but to a marsh.

25 Hudson Gorge Primitive Area

This protowilderness is too chopped up to be called a real wilderness, but its pieces, though small, have a provenance worthy of any Adirondack wilderness. Finch Pruyn's tracts split the part south of the Hudson River. The North Woods Club's holdings and access road border it on the north.

Of course, proximity to the Hudson meant that it was logged early, probably by the middle of the nineteenth century. The tract north of the Hudson that includes Kettle, Forks, and Pine Mountains was returned to the state by 1877. From the Blue Ledges east to the railroad near the confluence of the Hudson and the Boreas rivers, these mountains protect a deep gorge that could be called the Grand Canyon of the Adirondacks. The southern rim of this canyon, the Harris Rift Mountain and Fox Hill, was also part of the 1877 tax-sale lands. The forests closest to the Hudson tempted timber thieves until the turn of the century.

A fire swept up the western slopes of Kettle in 1908, but the rest was left to nature to restore, and among the restorations are one of the most spectacular cedar swamps in the north country. The huge, upturned stumps of fallen giants interwoven with trunks and twisted branches in a thick layer above the sphagnum-covered mire is impenetrable.

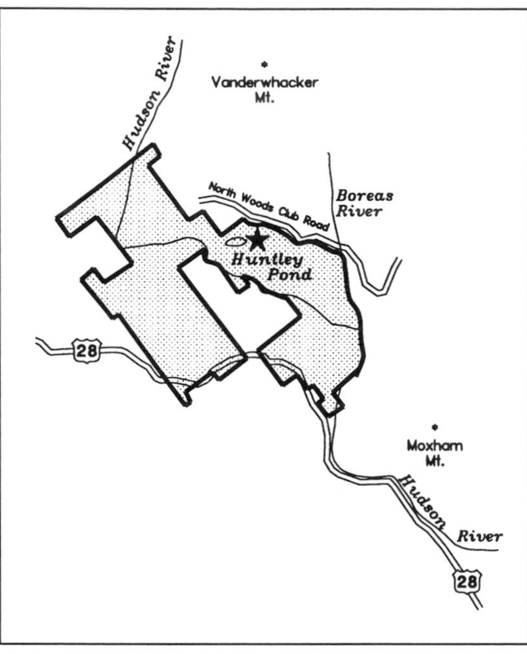

Lots south of the Hudson surrounding Big Bad Luck and Whortleberry Ponds were also returned to the state in 1877. Land south of the North Woods Club Road made its way into state ownership by 1888 because no one had paid taxes on this cut-over land. Finch-Pruyn Paper Company has owned tracts along the Hudson since the end of the nineteenth century. One of these tracts encompasses the confluence of the Hudson and Indian Rivers, which the company leases to the Gooley Club. The club posts that land, prohibiting public access, so that the public has limited access to the Hudson here.

Surprisingly few trails reach into this region, which is known best by rafters. Sportsmen's routes that can be adopted as trails by the state would increase access. Most marvelous to contemplate would be a rim trail along the northern flanks of the Hudson canyon. It could even be extended to Dutton Mountain in the Vanderwhacker Wild Forest.

Views and Visits

A spring raft trip, offered by many outfitters, is the best way at present to see the Hudson River Gorge. A drive along the North Woods Club to Huntley Pond hardly does justice to the region's forests. The short walk from Huntley Pond to the Hudson River Gorge traverses land logged more recently, but it does give a wonderful view of the cliffs of Blue Ledge.

The river itself is the highway for visiting the Hudson River Gorge. Here perky wooden McKenzie River boats guided by expert oarsmen challenge the river's rapids. Photograph courtesy of the author.

26 Vanderwhacker Mountain Wild Forest

Perhaps timber thieves and trespass explain why land surrounding Vanderwhacker Mountain did not receive wilderness classification. That tract, which is only part of this 70,000-acre wild forest, lies west of the Roosevelt Highway. It has one or two small, private inholdings, but it is a large enough area to be considered wilderness. Contiguous lots to the northwest of the mountain and scattered lots to the south of it became state land in the tax sale of 1877. These lots had been logged so early and were so close to navigable rivers (the Hudson and the Boreas) that their forests invited thieves in the late 1890s, and many logs were stolen from the edges of the region.

The Vanderwhacker Wild Forest encompasses land south to Moxham Mountain and east across the Roosevelt Highway toward Irishtown and Minerva. The borders of the eastern and southern portions are so mixed with private lands that wild forest status seems inevitable, but even here there are wilderness patches that have been undisturbed for over a century.

There were iron mines near Irishtown, a tannery in Minerva, and farming settlements throughout. The Boreas River, which cuts through the wild forest, was a highway for vast quantities of

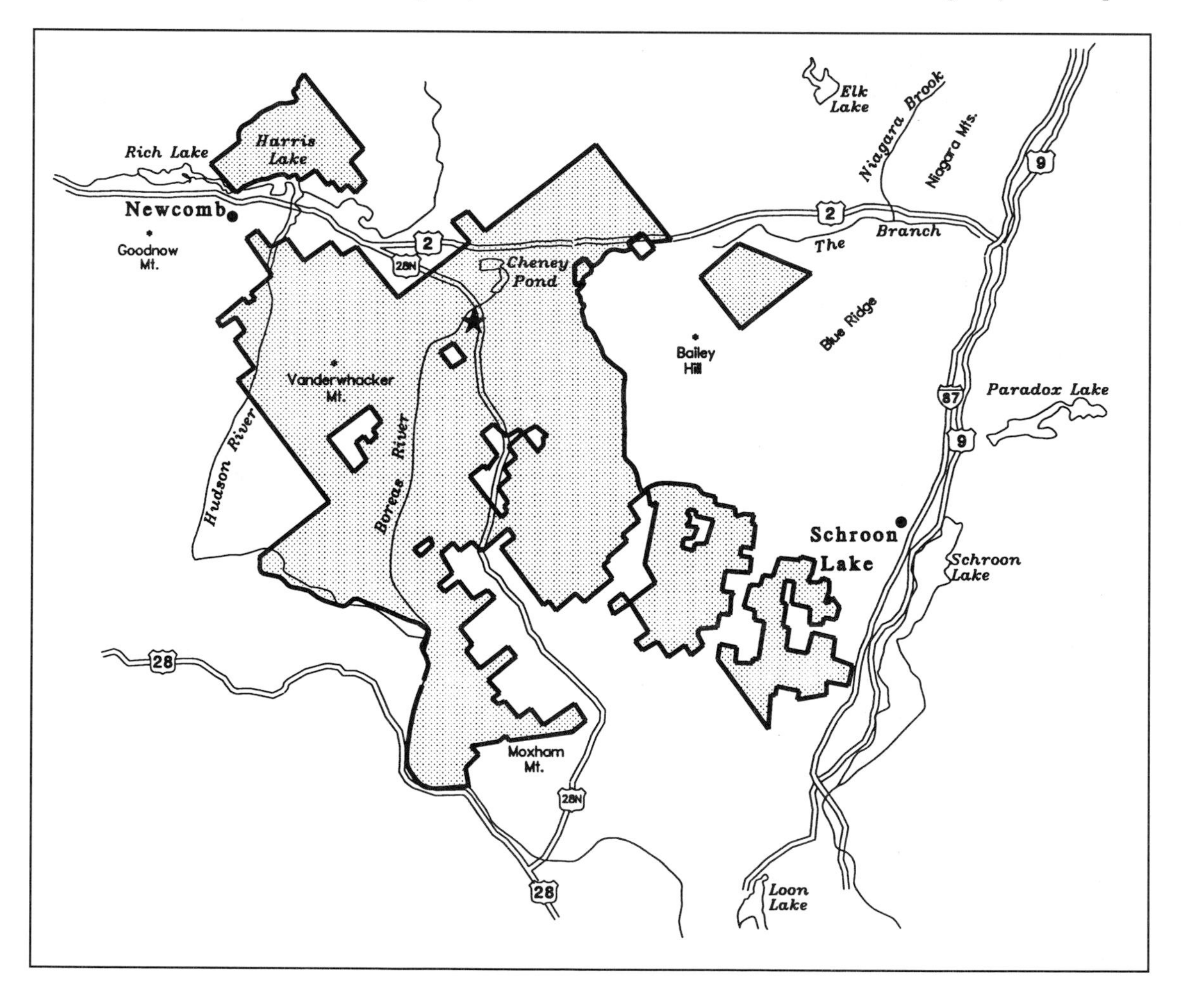

The Boreas River threads marshes, plunges through gorges, and here spills into a small waterfall above Hewitts Eddy.

Photograph courtesy of Wayne Virkler.

logs floated toward the Hudson. Dutton Mountain, with its spectacular views of the confluence of the Hudson and the Boreas rivers, is a part of this wild forest.

The Roosevelt Highway also cuts through the wild forest. It was named to commemorate Teddy Roosevelt's wild ride to the railroad station in North Creek so that he could assume the presidency after the assassination of President McKinley. The eastern border is a trail that follows an old road north from Irishtown along Minerva Stream, across the Boreas, to Cheney Pond.

Views and Visits

There is an easy trail to see the Boreas from N.Y. 28N. Alternatively, drive the North Woods Road to the Boreas and the railroad and walk along the tracks. Roads leading to the ponds east of the Roosevelt Highway are marked as snowmobile trails and are easy to walk. Cheney Pond Overlook south of the Blue Ridge Road is a picnic stop with a view south and southwest over the wild forest.

27 Hoffman Notch Wilderness

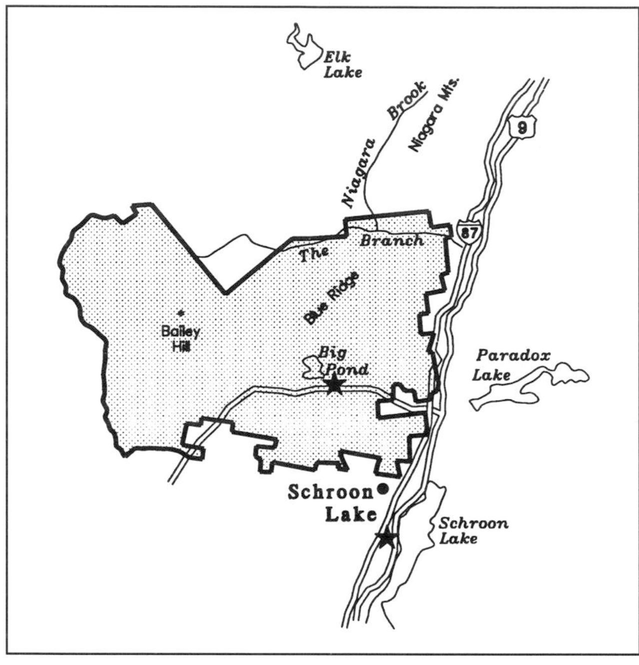

Hoffman Notch is bounded on the north by the old Carthage Road, on the west by Minerva Stream and a short stretch of the Boreas River, and on the south by Hoffman Road, a very old road that led west from Schroon Lake Village. Schroon River and Lake would define a fair barrier on the eastern boundary, but the parallel stretch of the Northway intervenes and makes the eastern border region visible but hard to reach. Hoffman Mountain, at 3700 feet, dominates this wilderness. In the confusing way the same name is given to several different places, the mountain complex containing Hoffman Mountain is called the Blue Ridge, while the Blue Ridge Wilderness is south of Blue Mountain Lake.

Old settlements along Hoffman Road and up through Loch Muller end at a trailhead for the Notch Trail, which divides the region north to south. Texas and Washburn ridges and Hornet Cobbles shelter a high valley filled with marshes through which the trail passes. Hoffman Notch Brook and the North Branch of Trout Brook drain those marshes. The trail through that pass has infrequent signs of long-gone settlers.

Big, North, and Bailey Ponds are gems in the southern portion of the wilderness. But the region's best-kept secrets lie in the eastern range containing Wyman and Squaw Mountains and the Peaked Hills. This range protects Hoffman's eastern flanks and makes that mountain almost unattainable. Capped with dense spruce and fir in impenetrable thickets, Hoffman, first depicted in Thomas Cole's 1838 painting, remains an artist's dream.

The central core of this wilderness was reacquired by the state for nonpayment of taxes in 1877, 1881, and 1885. The forests here are spectacular, with maples of enormous girth standing like sentinels along the southern portion of the Notch Trail.

The state's claim to lots in Totten and Crossfield Township 30, made in each succeeding tax sale year after 1877, was disputed. The state agreed to permit the disputants to log the land again in order to get clear titles to the land. Thus the northern part of the Hoffman Notch Wilderness was logged for small softwoods for pulp before its 1901 conveyance to the state. Its forests are still recovering.

The eastern slopes bordering the Schroon River had been logged and the valleys cleared for farming by sturdy settlers. Fires burned part of the top of Hoffman Mountain in 1903 and 1914, and patches south of Bailey Hill burned in 1905. Signs of old logging camps, abandoned stoves and broken equipment, and old roads survive. Only a few roads have become trails. The Blue Ridge Range and the hills to the east bordering on the Northway constitute a huge trailless area.

Exceptional forests grace some southern slopes, especially those north and west of Big Pond. There, deep valleys cut by the headwater streams of Trout Brook contain noble stands of hemlocks and hardwoods.

Like all wilderness areas, you have to travel toward the interior to find the best forests, but you can sample them on a walk past Big Pond toward the Hoffman Notch Trail.

The only way to reach the open rock crest of Jones Hill is to bushwhack around the shores of this beaver pond. Photograph courtesy of Chuck Bennett.

Views and Visits

Hoffman Mountain is visible from the Northway in the vicinity of Schroon Lake. Drive to Loch Muller and the farmhouse with the huge pine. It is private, but look across the fields to Hornet's Cobble and Hoffman Notch. For a short walk, try the easy and well-marked trail to Big Pond.

28 Hammond Pond Wild Forest

This wild forest was once so settled that it is astonishing that its core now seems so remote. The jagged hills and small mountains that rise sharply from Lake Champlain between Ticonderoga and Port Henry still echo the ring of forges and hammers. Iron mines, mills, brick beehives for making charcoal, roads, and finally railroads covered these hills in the nineteenth century. Forested hillsides were stripped bare for charcoal, and settlements sprang up at every mine and mill site—Hammondville, Ironville, Port Henry, Moriah, and Witherbee.

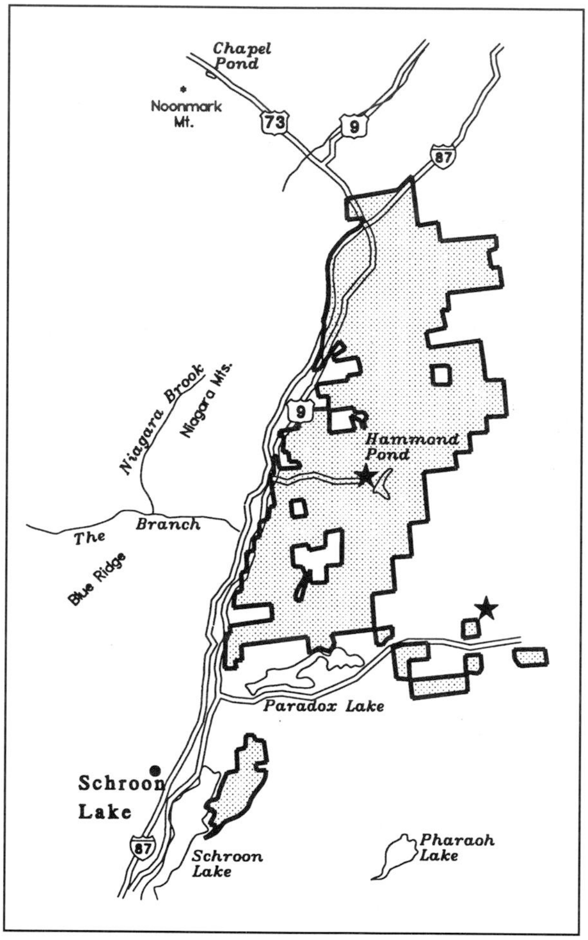

The first road to penetrate the mountain core from Lake Champlain crossed the northern part of the region. The Cedar Point Road, built in 1828, headed west from Port Henry, then curved south through a mountain pass to the valley of the Schroon River. There it began to climb another mountain pass to the Valley of Niagara Brook. The stretch of road beyond Niagara Brook and west past Clear Pond was part of the original route to the mines at Tahawus.

The most important east-west road in the Adirondacks crossed the southern part of the region. The Carthage Road, dubbed the Catamount Road, began at Crown Point and wandered west past Ironville and Johnson Pond to Roots' Hotel on the Schroon. From there it headed almost due west toward Tahawus and Newcomb.

These two highways of the past were incredibly difficult roads, rough with corduroy, steep, and rocky. That they penetrated the mountains west of Lake Champlain was a feat; that they found passes through the even more rugged mountains to the east of the Schroon was a testament to those who would seek Adirondack ores and forests.

Roads connected every settlement close to Lake Champlain. Few settlers headed into the western mountains closer to the Schroon. The eastern part of the region is still settled—all the original towns except Hammondville survive. The western part is the core of the Hammond Pond Wild Forest, and only the settlments along the Schroon River survive.

As the mines began to fail and the forests were stripped, owners of the taller mountains forfeited their worthless lands, beginning in 1871. The state continued acquiring land through tax sales, but at a fairly slow rate. Even when the state began to purchase land in the future wild forest, acquisition proceeded slowly through the first part of the twentieth century, so that as late as 1920, the state owned little more than half the 40,000 acres that now make up the wild forest.

Although the wild forest is laced with roads, few old routes survive as trails, and many that do are blocked by tracts that are still privately owned. Two networks—one connecting Bass and

Bloody Mountain overlooks Hammond Pond. From its open summit you can see Nippletop, one of two mountains so named in the Park. The views are great in both summer and winter.
Photograph courtesy of Chuck Bennett.

You feel almost as if you could leap into Arnold Pond from Skiff Mountain. It is so steep it provided the perfect locale for a kestrel to teach its young to fly and hunt. Such serendipitous views of nature enhance mountain scenes in the Hammond Pond Wild Forest. Photograph courtesy of Wayne Virkler.

Berrymill and Moose Mountain Ponds, the other leading out from Sharp Bridge Campground to Moriah and Crowfoot Ponds—constitute almost all the state trails. Even though old roads abound, they just do not lead to the wonderful open mountains that have some of the best views in the Adirondacks. Hail, Bloody, Fern, and Moose Mountains and Harris Hill, Bald Pate, and Owl Pate all have wonderful views, their steep slopes bared by the turn-of-century fires that swept virtually every peak. The triangle between the road along Black Brook and the Moriah Road that encompasses Hail Mountain, then known as Blue Ridge, was called waste and denuded land in 1908. There was nothing left there to burn.

Views and Visits

Visit the Hammond Museum in Ironville, or walk the short trail from County Route 4 to Hammond Pond.

29 Dix Mountain Wilderness

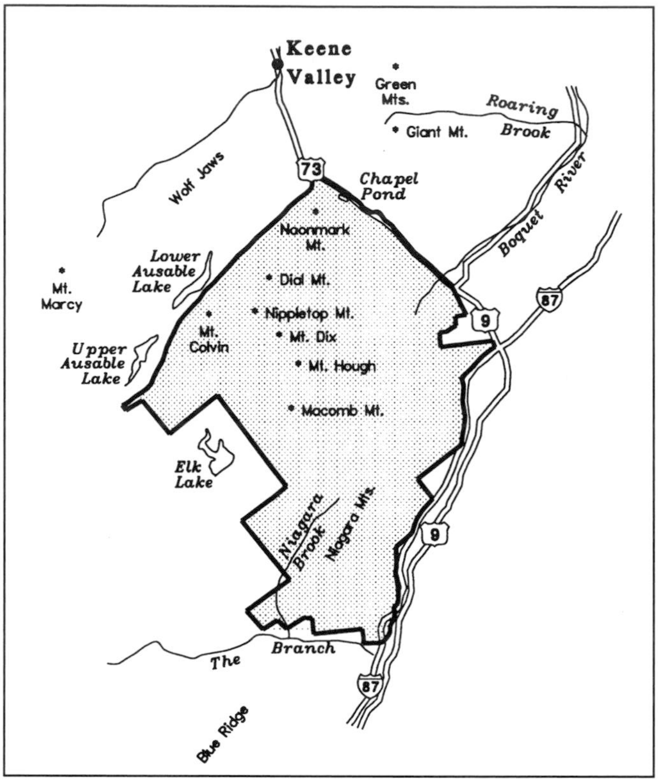

Fires ravaged the Dix Mountain Wilderness Area in 1903, 1908, and 1913. Before the turn of the century, its lower slopes had been logged for spruce and balsam, and the slash from logging here and on adjacent mountainsides provided a tinder box in those dry years. In the wake of the fires, pioneering paper birch seeded through the brambles, and gradually a coat of pale green cloaked the cirques and ravines. White pines found footing on dry patches where only lichens first appeared on the burned soils.

The lighter green of paper birch is still dominant, but the birches are approaching the end of their short life spans. Nowhere in the Adirondacks are there more beautiful and stately paper birch forests, but all that will shortly change. The lighter green shades of this trapezoidal tract are giving way to the darker tones of spruce and fir.

Beginning in 1898, the state began to purchase lots in Township 49 from the Adirondack Mountain Reserve (AMR). The state obtained the summit of Nippletop in 1933, and in 1978 the tops of Blake, Colvin, the Dials, and Bear Den, summits that stand above the northern border of this wilderness. The border also includes Pinnacle, Nippletop, and Noonmark, which are suspended above the deep trench of the Ausable Lakes and River. Private land around the Ausables severs this wilderness from the High Peaks.

The massive complex of Dix itself, with Hough, Macomb, and East and South Dix, lies near the western boundary that abuts the private lands around Elk Lake. The Northway and the highway through the Chapel Pond pass forms the eastern boundaries.

All this background does not begin to explain why this is the most beautiful and my favorite of all the northern wildernesses. Perhaps it is the attraction of the many forks and branches that collect on the east to form the headwaters of the Boquet. The south fork drains the cirque between Dix on the north and the long range of Spotted, Elizabethtown Number 4, and East Dix on the south. Sparkling streams with their secret waterfalls contrast sharply with the torrents that sometimes scour the white boulders of the stream beds. These torrents occur when raging storms soak the mountainsides and peal away trees, rocks, and soil to leave bare scars. Occasionally such slashes start high on the mountainsides.

Farther south, the quieter valleys of Lindsay, West Mill, and Walker Brooks drain into the Schroon River. These streams pierce a little-known mountain range with sharp trailless peaks. Walker Brook comes from a high cirque between Niagara, Nippletop (there are two mountains with this name), and Camel Hump. Niagara Brook rises high on the slopes of Sunrise and, with a waterfall, plunges to the valley below before turning south toward The Branch. The headwaters of West Mill Brook, rising on the slopes of Macomb, also drop in a hidden waterfall, which few have seen.

Rugged, jagged, scarred, the mountains visible from Dix make up one of the Adirondack Park's most spectacular ranges.

Photograph courtesy of John E. Winkler.

Every one of these peaks has a rocky, open summit. Only the Dixes themselves and the northern range above the Ausable Lakes have trails. There is a lifetime here of hidden places to explore, accompanied by the isolation and quiet that give the feeling of solitude that makes it all worthwhile. The only drawback is the Northway, which does protect the southeastern boundary (access is only through covert underpasses) but can send its traffic noises to the summits of the border mountains if the wind is from the east.

What appeals to me most are the open slopes and summits—the chance to wander from one outcrop to the next and to enjoy the ever-changing vistas. And for this we have to credit the fires for clearing away the dense forests, revealing scarps that are so steep no forest can gain a foothold to block the views.

Views and Visits

There is no easy way to sample this wilderness. From the Northway near Schroon Lake you get a fleeting glimpse of Dix with its prominent knob, the Beckhorn, and longer views of the knobs of Macomb, Hough, and South Dix. Farther north you pass mountains of the eastern border range, but there is no good place to stop and identify the peaks. For mountain climbers, the view of the Dix Range from Noonmark is the best way to sample the range.

30 Giant Mountain Wilderness

The Giant Mountain Wilderness Area appears as a massive rock core with smaller peaks on the northwest. The center of the block, the heart of the Roaring Brook Tract, some 6,000 acres, became state land as it was abandoned, mostly in the 1870s. Some lots have always been state land. The bare granite summits were always forbidding; no harvestable forest covered the upper slopes. Climbing from the east and the Boquet Valley south of New Russia, a trail traces the knobby summits and outcrops of Rocky Peak Ridge over to the slopes of Giant. Here the soil, what there is, is thin, and the trees on upper slopes are scrubby. Parts were always considered waste and denuded by early surveyors, even on their earliest maps. No wonder the fires of 1903 and particularly those of 1913 swept over the slopes of Giant, Rocky Peak Ridge, and north to Green Mountain and Knob Lock. The virgin hemlock forests on the lower slopes of Giant's northwest face barely escaped burning.

What the fires did not take, the pulpers did. Baxter and other smaller peaks east of Keene Valley were logged by the early 1920s, at a time when the Adirondacks was running out of spruce and fir. Only when these lower hills were stripped of their pulp logs did their summits become state land.

Wind and water finished what fires had started. Giant's cirque is scarred with slides, the largest of which in recent time was the slide of June 1963, which washed stark piles of rubble all the way to N.Y. 73. The detritus temporarily changed the path of the stream that flows over Roaring Brook Falls.

Views and Visits

The summit of Giant and Rocky Peak Ridge loom above surrounding roads and are visible from several vantages. Roaring Brook Falls, the course of its waters restored, is visible from N.Y. 73. The best view of Giant is across the private golf course at the Ausable Club, on grounds belonging to AMR. Since there is no parking on AMR roads, you have to walk the road from the parking area to enjoy this view.

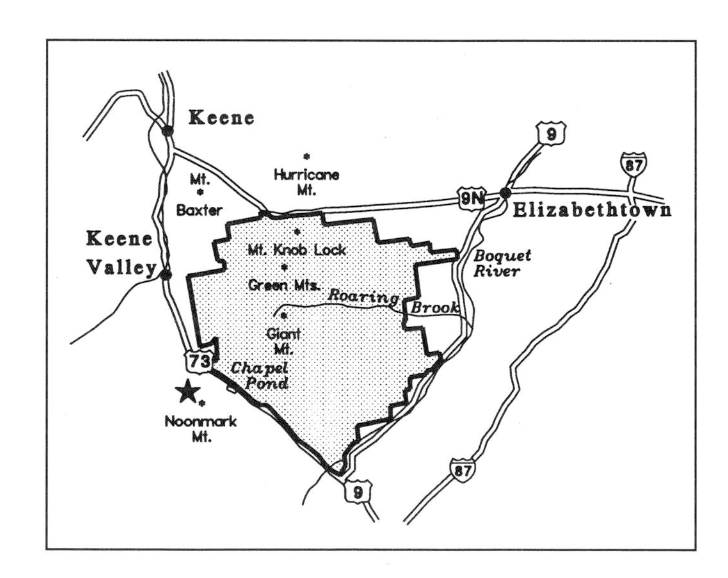

No wonder the ridge stretching away in this picture is called Rocky Peak Ridge. It is typical of the scarred and open rock-faced mountains of the Giant Wilderness. Photograph courtesy of Charles H. Bennett.

31 Hurricane Mountain Primitive Area

From the summit of Hurricane Mountain north, all this primitive area burned in 1906. A few lots on the steep south side of the mountain are preexisting Forest Preserve, never logged and never burned. A network of lovely trails—over the Crows, approaching Hurricane itself from three directions, and around the cirque backed up by the Soda Range and Weston Mountain on the Nun-de-ga-o trail—make this eminently accessible.

The fire tower on Hurricane keeps this area from being a designated wilderness. Vehicular use along Luke Glen Road keeps it from being combined with the Jay Wilderness to the north. But these are mere technicalities for a wild area with such easy trails.

One of the delights of the Hurricane Wilderness is the hike up over the Crows.

Photograph © 1998 by Carl E. Heilman II.

Views and Visits

None of the trails mentioned above is easy, but the trail to the Crows is suitable for beginners.

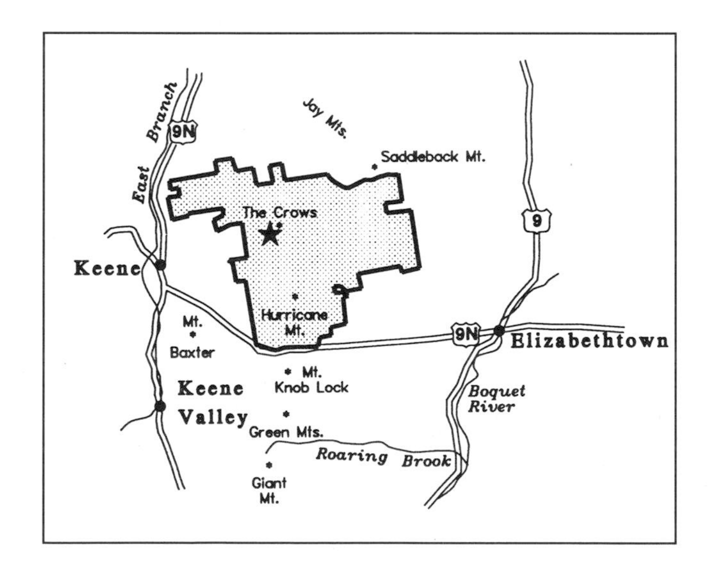

32 Jay Mountain Wilderness

Jay Mountain has been designated a wilderness area even though its near-twin region to the south retains its primitive designation. The two regions share a common and fiery history— most of their surface was burned in 1906 fires. The legacy of the fires is the wild and forsaken appearance of the long, thin, curved ridge that supports Jay's summits. Bare rock, lichen patches, stunted shrubs, and fields of scree give an aura of desolation mixed with strange geometric forms. This feeling of desolation is enhanced by a series of tall cairns, rocks piled high by unknown hikers to stand like ghostly sentinels along the ridge line. Infinitely more picturesque than ordinary wooded mountaintops, Jay would be an attractive destination even without the spectacular views it offers. And the unique and open chain of knobs has views from Whiteface to Vermont.

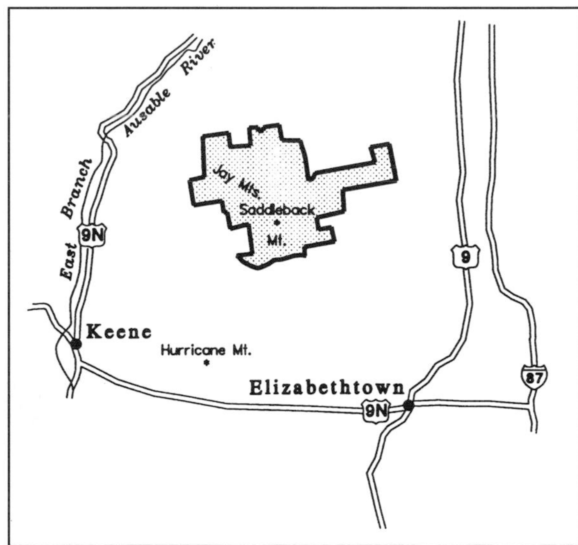

Even the second-growth forests growing on its slopes—stands of red pine, thickets of paper birch, scattered hardwoods that are reaching maturity—are special, and pockets of tiny marshes quilt the slopes up to the dense spruce-fir thickets that protect the summit ridge.

A ring of mountains circling the eastern border of the wilderness is as inaccessible as Jay itself. Saddleback, Slip, Seventy, Death, and Arnold Mountains and Bald Peak all have summits as bare as Jay; none has trails, and all are treasures of vantage points, cliffs, and stunted trees.

This wilderness is almost completely surrounded by private lands, and there are no state trails here. The steep slopes are ramparts protecting a keep accessible only to those who need no trails.

An iced landscape on the summit of Saddleback in the Jay Range is the reward of a winter climb. Whiteface is in the background.

Photograph courtesy of Charles H. Bennett.

33 High Peaks Wilderness

The High Peaks Wilderness is really two regions, split by Indian Pass. The smaller, eastern core is the most well known, most visited hiking destination in the Adirondacks, perhaps in the whole northeast. Its peaks, from the state's tallest (Mount Marcy), include twenty-six of the forty-six High Peaks, those over 4,000 feet in elevation. Its southeastern border encompasses the Great Range, which rises above Johns Brook Valley and the valley of the Ausable Lakes and river. Private lands around the Ausable Lakes are the actual southern border.

The largest patch of land in the Adirondacks that has never been privately owned is southwest of Street and Nye Mountains. Other preexisting lots, or those that have always been owned by the state, are scattered among lots that have been continuously owned by the state since 1877 and probably have never been logged. Lots west of the Loj Road and south of Raybrook, all the way to Wallface, with one exception, were neither logged nor burned.

In contrast, land immediately around the Loj Road, and the dagger to the south, was severely burned in 1903 in a blaze so severe it is one of the few where fire destroyed virgin forest.

Lands in the south around the iron mine at Tahawus remain private but were once much more extensive. When the supply of spruce had dwindled in all the rest of the park, spruce and balsam fir were sought on the steepest slopes. Finch Pruyn and Co. logged from Wallface south, along the slopes of the MacIntyre Range, and around Flowed Land and Lake Colden, just before 1913.

The High Peaks has always been the province of hikers and adventurers from the first group to seek the state's highest mountain in 1837 to the more than 100,000 hikers who visit annually. Adirondack guides hacked out the first trails through the impenetrable spruce-fir thickets that surround the highest peaks. A half dozen of those peaks are naturally bare, rising above the tree line, with only fragile Alpine flora clinging to crevices. From the 1860s on, guides like John Cheney, Orson Phelps, and Bill Nye built trails straight to the summits. Later, hiking groups added more trails until the eastern High Peaks were covered in the densest network imaginable.

Oddly enough, many of the highest summits did not belong to the state, but to the Adirondack Mountain Reserve. In a series of purchases between 1921 and 1932, the state obtained the northern slopes of the Great Range from the western slopes of Mount Marcy to Lower Wolf Jaws. There trails up the south slopes had been built by the Adirondack Trail Improvement Society, an independent organization started by members of AMR but open to the public.

The plethora of trails in the eastern High Peaks, plus incursions of private land and roads and the state's inability to impose reasonable limits on access to the region make it difficult to see why this area is called a wilderness. This is additional proof that wildernesses just do not happen, they are planned. It also affirms the contention that poorly managed wildernesses soon lose their wilderness flavor.

In the western portion of the High Peaks the mountain ranges are broken by long valleys, the longest of which surrounds the Cold River. These valleys invited roads and trails. The flat terrain makes the valleys and flatlands horse-travel country. The 1950 blowdown was especially severe in the Sewards and not only was the blowdown salvaged (cut up and hauled away), roads were built to help the salvage and ostensibly to provide future fire protection. That explains why many of the trails seem more like roads even today.

The Seward Range, with Seymour and Donaldson, has no trails, although visited by mountain climbers over established routes. Only the adventurous seek the other routes. The Sawtooth Range—jagged enough to live up to its name—and Cold Brook Mountain are almost never visited.

Private lands, some owned by Finch Pruyn, some part of Huntington Preserve, a research station belonging to the College of Environmental Sciences and Forestry at Syracuse, make up the southern border. Santanoni and its adjacent peaks are nominally trailless, but the approaches are along well-worn old logging roads. The Preston Ponds, two gems near the headwaters of the Cold River, and the land surrounding them are high on the list of future state acquisitions.

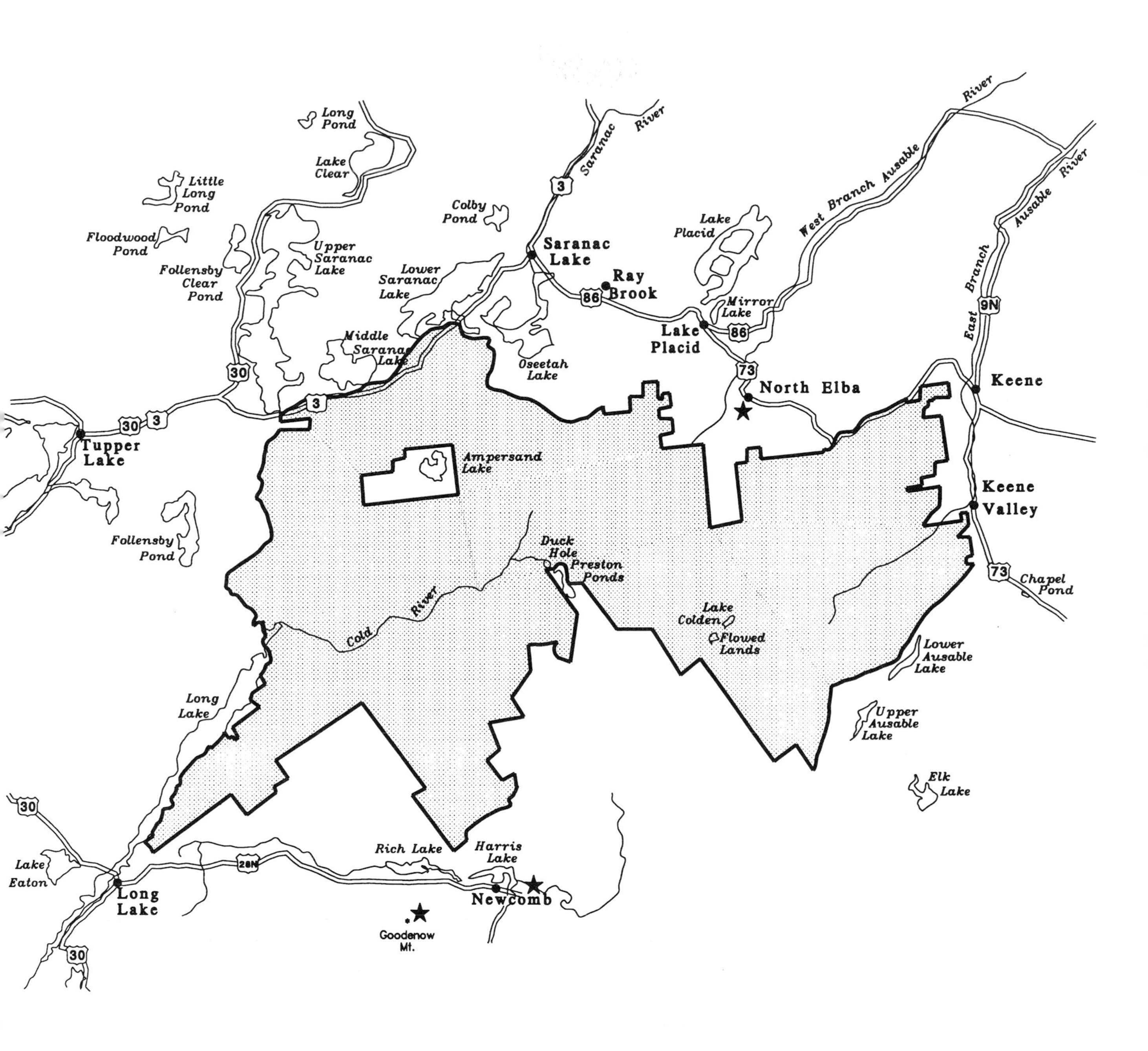
Long Pond
Lake Clear
Little Long Pond
Floodwood Pond
Follensby Clear Pond
Upper Saranac Lake
Saranac River
Colby Pond
Saranac Lake
Lower Saranac Lake
Ray Brook
Lake Placid
Mirror Lake
Lake Placid
West Branch Ausable River
East Branch Ausable River
Middle Saranac Lake
Oseetah Lake
North Elba
Keene
Tupper Lake
Ampersand Lake
Keene Valley
Follensby Pond
Duck Hole
Preston Ponds
Chapel Pond
Cold River
Lake Colden
Flowed Lands
Lower Ausable Lake
Long Lake
Upper Ausable Lake
Elk Lake
Rich Lake
Harris Lake
Lake Eaton
Long Lake
Newcomb
Goodenow Mt.
30
3
86
73
9N
28N

While hordes of people detract from the wilderness character of the eastern High Peaks, the broad, roadlike trails detract from the wilderness character of the western High Peaks. To find the wilderness you have to step off the trails and bushwhack to Blueberry Mountain overlooking Long Lake. The truck trail (limited to hiking) from Coreys to Duck Hole is a virtual highway, but its surroundings hold many secret places.

What makes this place a wilderness? The forests around Ampersand have never been logged. Forests west of Long Lake were logged but only sparingly for large softwoods. In the 1870s, there were plans to divert the waters of Long Lake through a canal to Round Pond and thence by a chain of lakes to Rich Lake and the Fishing Brook source of the Hudson. That would have diverted water from the Raquette River to the Hudson to enhance spring floods, which was necessary to drive logs down the Hudson River. If the waters had been diverted, it is certain that much more of the Raquette River watershed would have been logged more heavily.

More important than any past history is the huge size of this wilderness, the largest single block of state land in the Adirondacks. With a few acquisitions to round out the borders and stronger policies to redirect recreation and limit the numbers of hikers in the most overused places, this area could unquestionably earn its title as the most important wilderness area in the eastern United States.

The slide on Santanoni is the longest in the Adirondacks. It lies deep in the southern edge of the High Peaks Wilderness. Photograph courtesy of the author.

View from the summit of Algonquin in the MacIntyre Range. This summit is one of the most popular in the High Peaks Wilderness.

Photograph courtesy of Tom Bessette.

Views and Visits

There are two very famous views of the High Peaks—one across the fields at North Elba, just west of the Loj Road, looking south to MacIntyre and the cut between it and Wallface, the other looking north from the picnic ground near the Town Office in Newcomb. The well-designed trail up Goodnow is suitable for beginners, and its view north to the High Peaks is spectacular.

34 Sentinel Range Wilderness

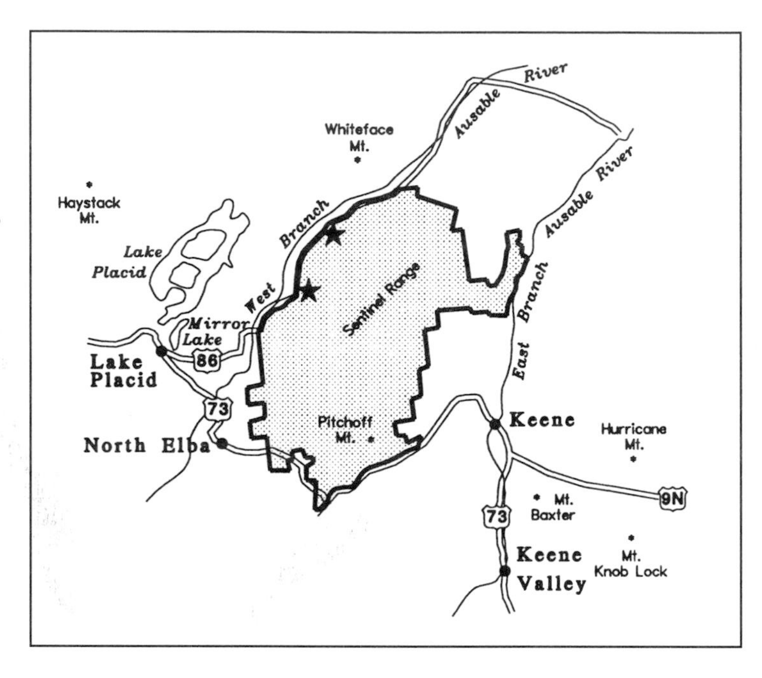

Few wilderness areas in the Adirondacks have a smaller proportion of trails than does the 24,000-acre Sentinel Range Wilderness. And, not surprisingly, both geography and history play a role in this dearth of trails. The mountainous core has no streams capable of floating logs, although the Ausable River on the western boundary was a highway for logs. Even though the range is visible from many vantages in Lake Placid and surroundings, lumbermen must not have coveted its slopes or perhaps were put off by their formidable contours. A few lots were never sold by the state—they constitute some of the small acreage in the Park that has never been in private hands.

Nearly three times the acreage of those original lots are in surrounding tracts that were forfeited for taxes in 1877 and never reclaimed. Among these lots were tracts bought by Peter Smith. His son, Gerrit, gave part of the large tracts of land he inherited from his father to members of the black colony he tried to establish around 1846. Among those who received land from Gerrit Smith was the abolitionist John Brown, who settled in North Elba. The surveyor of one such gift lot considered it "hardly worth paying taxes on." Hence, much of the mountainous heart of this wilderness was never logged because by the first decades of the twentieth century, when lumbermen had run out of spruce in lower elevations and were looking to forests on the higher slopes, the core of the Sentinel Range had been safe in the Forest Preserve for many years.

The northern border of the wilderness, the Whiteface Tract portion, was logged by J. and J. Rogers Company. Traces of those logging roads are still visible, but few of the mountainous routes have ever been marked as trails.

Cliff-faced mountains rise above both sides of the Ausable. On the west, in this wilderness, secreted in clefts at the top of this range, are tiny pond, Owens, Winch, and Copperas, all joined by a fine network of trails. Parts of the western border of the wilderness was logged during the first two decades of the twentieth century. Farther south, the tops of these mountains were logged probably during the same time from access roads that later became the North and South Notch ski trails. These trails, too, are now gone.

In 1908, fires swept over all of Pitchoff Mountain, burning it so severely that its open knobs and summits are still bare. That fire swept down the northwest flanks of Pitchoff to an old road, a part of the Northwest Bay Road whose origins around 1810 led to its being erroneously labeled "Military Road." It wound through the valley north of Pitchoff and along Nichols Creek. Today a popular ski trail follows that route. Trails to a cluster of tiny ponds in the northwest corner, close to the Ausable Valley, constitute the rest of the region's marked routes.

Like most of the Adirondack's wilderness regions, the Sentinel Range was built around tracts that were almost inaccessible and at critical times in their history considered worthless.

Nature has conspired to build a trail

Nature's trails are always a challenge. This slide on Kilburn in the Sentinal Range is no exception. McKenzie Mountain and its wilderness area are in the background. Photograph courtesy of John E. Winkler.

where none ever existed. The steep west-facing slopes of Kilburn, one of the giants in the Sentinel Range, were ravaged by a slide that occurred in a violent rainstorm in the fall of 1994. That natural scar faces Whiteface and is visible from many places.

Views and Visits

Drive along N.Y. 86 for glimpses of the wilderness. For the very adventurous, walk along an old logging road that leads into the woods opposite Monument Falls on the Ausable. A well-worn path following the old roadway leads in thirty minutes to a great mass of broken and twisted trees at the base of the Kilburn slide. Turn south here and head up along the stream bed and through the rubble to view the great slide.

35 Saranac Lakes Wild Forest

With 68,000 acres, this wild forest is large enough to qualify as wilderness, but the Saranac Lakes Wild Forest, which encircles the lakes and the private lands bordering them, is broken into several disjoint parcels, each distinctly different.

The hills north of Middle Saranac were heavily logged, but touched only by isolated fires. The region west of Upper Saranac was logged by the forester Bernard Fernow, who wished to prove that scientific forestry could succeed. His plans were undone by the fires that swept over the tract in 1908, but the recovered forests and reforestation areas shelter some of the loveliest hiking and skiing trails in the park.

A patch of land south of Ray Brook is also included in the Saranac Lake Wild Forest. Land adjacent to the railroad burned with such severity that reforestation was necessary. The handsome trail up Scarface Mountain begins in these new forests, which are fast reaching maturity. A trail up neighboring Seymour Mountain is definitely needed—this summit offers an easy climb with a range of views comparable to Scarface.

This wild forest is best known, however, for its waterways and islands, its campsites and secluded bays. Despite the fragmented nature of the wild forest, it is a model for the way new forests can create a wilderness.

Views and Visits

Walk the spectacularly varied reforestation trail in Fernow Forest. Gaze up at the cliffs on Scarface from the highway near Ray Brook. Sample the reforestation areas along the trail to that mountain.

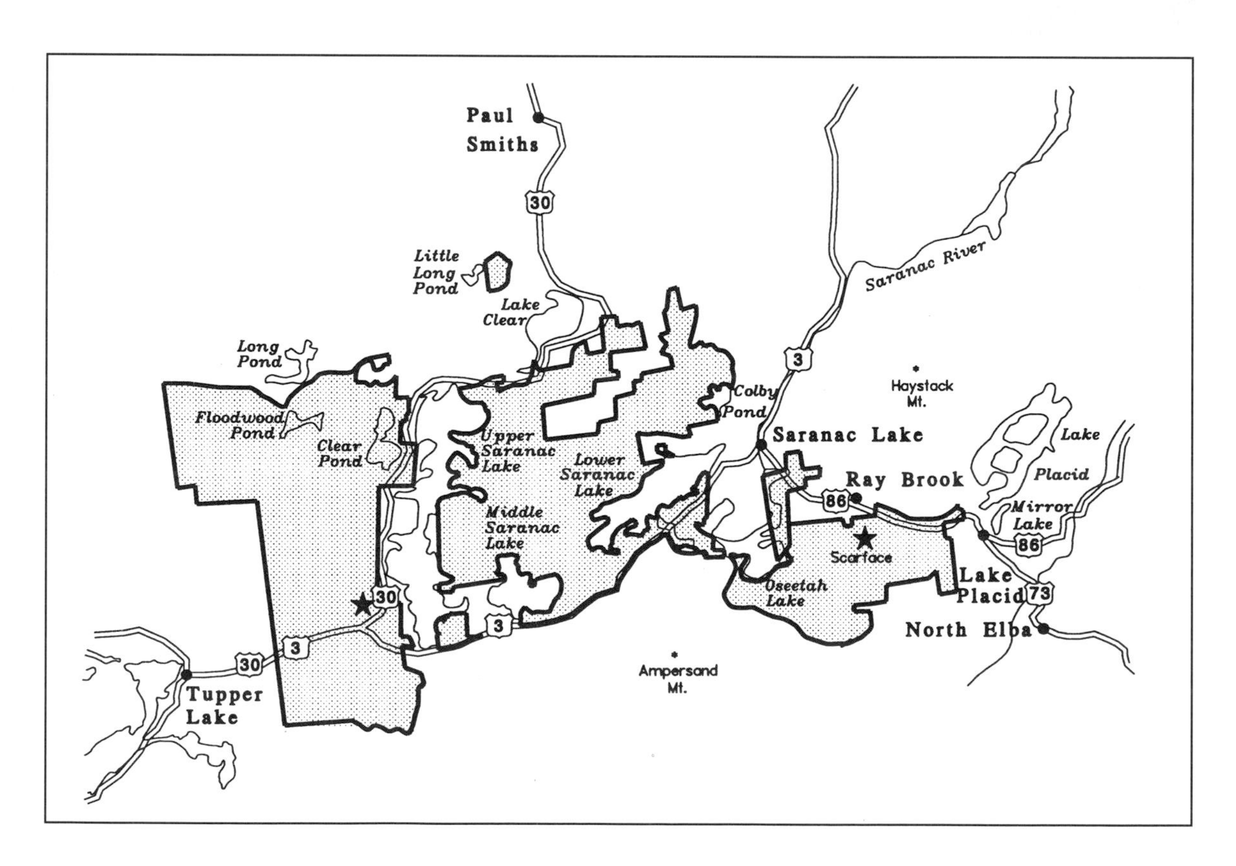

Huge cliffs visible from the highway give Scarface its name. This small mountain has views toward the Saranac Lakes as well as the High Peaks.

Photograph courtesy of John E. Winkler.

36 McKenzie Mountain Wilderness

At 40,000 acres, the McKenzie Mountain Wilderness is a surprisingly large wilderness area, given its settled surroundings: Lake Placid and Saranac Lake on the south and the Whiteface complex of Memorial Highway and ski center on the east. The Saranac River defines its western and northern boundaries, and views across the valley of that river offer glimpses of a wilderness you almost feel you can touch. What is amazing about this wilderness is how little of it is accessible.

The provenance of the wilderness core has much in common with other wilderness areas. As much as half of its lots reverted to the state for nonpayment of taxes before 1877. Because no logging was done on the higher Adirondack mountains before that date, it is likely that almost no logging ever was done on these lots. They encompass the north flanks of McKenzie Mountain, all of Moose Mountain, and west to the summits of the small range that includes Blue Mountain, Owls Head, Pigeon Roost, and Mount Alton. The latter four names are unfamiliar because they are almost never visited. There are no logging roads in the interior, and no interior rivers on which to float logs except the Saranac, which lies at the base of this isolated massif.

The 1908 fires were able to touch the wilderness, but only barely, on its southern hills. The fires, which started by the railroad extension to Lake Placid, swept over the summit of Haystack and up the south slopes of McKenzie, scarcely invading this mountain fortress. Trails now follow the fire route up the slopes of Haystack to McKenzie.

Summits capped with dark spruce-fir stands and lowlands of mature hardwoods contrast with the brilliant trunks of paper birch that mark the disturbed areas. The mountains are a pristine watershed, draining into Lake Placid and providing pure water that needs little treatment for the communities of Saranac Lake and Lake Placid.

The northwestern corner of the wilderness is another story altogether. The slopes of Whiteface were logged in this century by the J. and J. Rogers Company. They sought pulp woods for the paper mills that replaced that com-

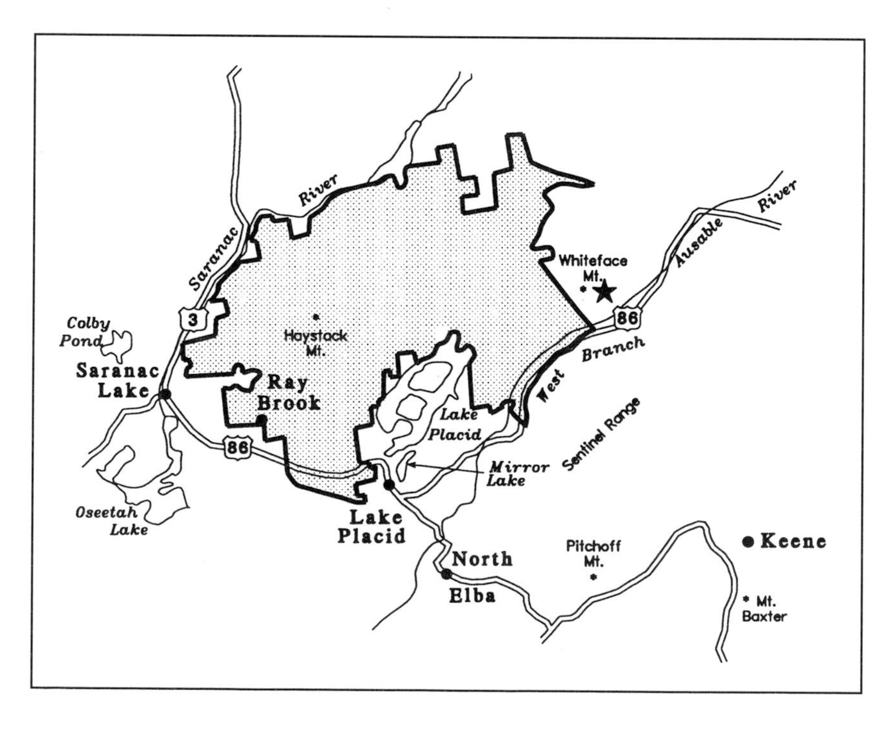

pany's iron mills. The logging was extensive, with logs floated down long flumes that led to the Ausable River. That one logging, which ended before 1920, took even small softwoods, but the forest is recovering. In places there is an unnatural preponderance of hardwoods, but there is something about a single logging episode that favors forest regrowth.

Today, the contrast between the disturbed and untouched is fading. The nineteenth-century adjectives—inaccessible, rugged, "worthless"—are now equated with such terms as protected, intact, and inviolable.

Views and Visits

Views from Lake Placid and the highways surrounding the wilderness do little to reveal the wilderness, nor does a drive up Whiteface Mountain Highway, but that drive offers a beautiful panorama of the untouched core across Lake Placid.

Deep woods, carpeted in tall ferns, border the trail to McKenzie Mountain.

Photograph courtesy of the author.

37 St. Regis Canoe Area

All you need for a great canoe area is a cluster of connected ponds with carries of varying lengths between the ponds. Judging from the popularity of the St. Regis area, it would really be fantastic if the Adirondacks had a couple more canoe areas. The trouble is—there are just no other comparable areas.

The best candidates are private. Whitney Park has enough ponds connected by roads that could be used for carries to link the existing canoe routes on the Raquette with headwaters of the Beaver River and the Bog River, using several different pathways. All the ponds that flow into the headwaters of Black Creek are on state land, except privately owned Jerseyfield. Having access to that lake would make a small, canoeable cluster, but nothing like the St. Regis area. The wilderness ponds north of Stillwater are a possibility, but the terrain between ponds is probably too difficult to make a canoe circuit.

Great campsites, lovely, wooded trails, a mountain to climb (Long Pond Mountain with its new trail), great forests, and relatively level terrain make this area perfect for canoeing. And it probably cannot be duplicated.

Some of this forest was logged in the 1860s, but it remained so remote that it was one of the last refuges for the almost extirpated beaver. All 26,880 acres of Township 20 of Macomb Purchase Great Tract 1 was purchased by the Upper Saranac Association and logged again in 1886, but they never cut trees close to the shores of the lakes and ponds. What is now the St. Regis Canoe Area was purchased by the state in 1898 and never logged again. This kind of preservation guarantees the pristinely beautiful wilderness setting for these lakes.

Views and Visits

Drive along Floodwood Road and enjoy the magnificent forest, or walk the short canoe carry to Long Pond for views of the area across the pond.

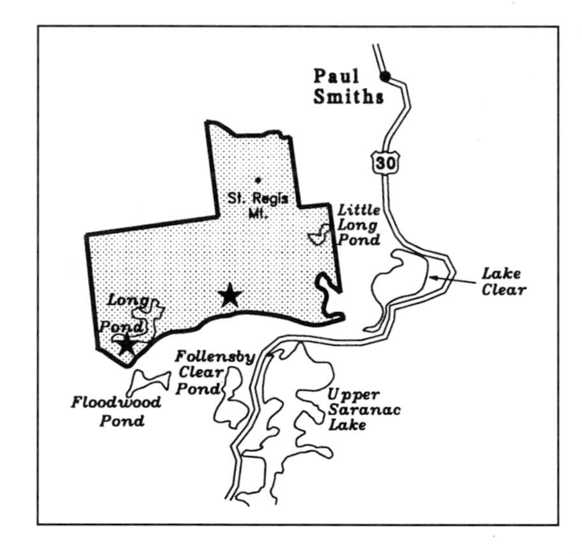

The magnificent reflections on blue water await those who are lucky enough to find a calm day to canoe into the St. Regis Canoe Area.

38 DeBar Mountain Wild Forest

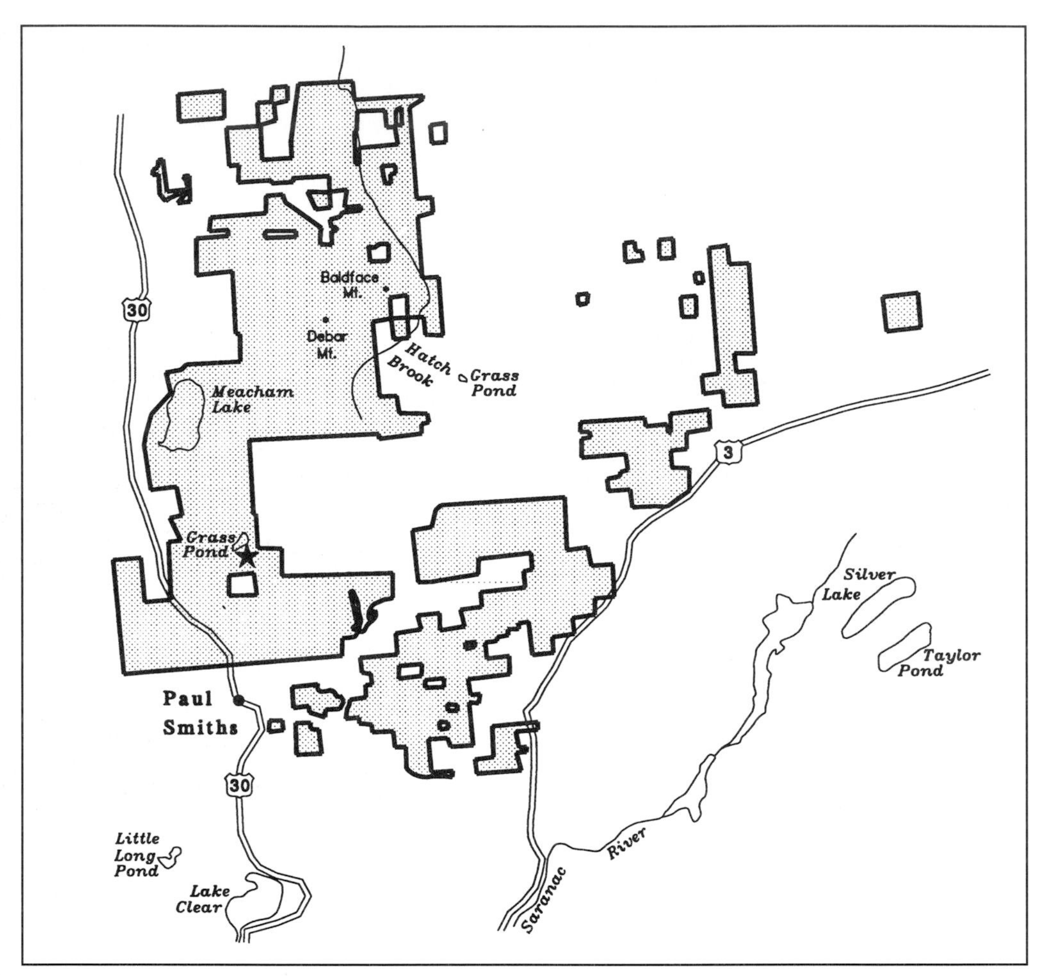

Fire has marked huge swaths of the 83,000-acre DeBar Mountain Wild Forest—the intense fires of 1908 swept through the heart of the rectangle that reaches from Osgood Pond north past DeBar Mountain. That isolated peak, a former fire-tower mountain, is the sentinel of the northern Adirondacks, overlooking the land as it flattens out in the plains along the St. Lawrence River.

Meacham Lake lies at the foot of DeBar Mountain, with a state campground at its northern shore. This large lake once had two hotels, but both burned long ago. Most of the trails in the wild forest emanate from the campground or from the Hays Brook trailhead that is north of Paul Smiths. Long, flat trails skirting wetlands around Hays and Hatch Brooks cut diagonally across the wild forest.

This land was once renowned for its game and for the guides who helped sports find its deer, moose, panthers, and wolves. Later it was known for an attempt to colonize its fields with elk. Currently, the ponds and wetlands created to encourage game animals are maintained only by nature, which is slowly returning moose to the region.

The core of recreational activities in the area deserves expansion to include trails on Baldface Mountain with its double summits and extraordinary panorama of the northern Adirondacks.

Views and Visits

Walk to Grass Pond on a route that was once through waste and denuded lands.

The cliffs on Jenkins rise above a beaver pond. The wonderful new trail follows an esker that creates the pond.

Photograph courtesy of the author.

39 Silver Lake and Taylor Pond

Appended to the isolated complex of state land encompassing the south shores of Silver Lake, all of Taylor Pond, and Catamount Mountain is a small patch of state land that also includes the summit of Silver Lake Mountain. A great beginning has been made in the development of the recreational potential of this area. The trail up Catamount may soon have a trailhead worthy of such a great, rocky mountain. The Nature Conservancy has built a fantastic boardwalk and trail through its preserve at Silver Lake Bog and the ridge overlooking Silver Lake.

Trails up the north slopes of Catamount from Taylor Pond are needed. Taylor Pond has a very nice campground and a wonderful circumnavigating trail.

Silver Lake Mountain has a new trailhead. Acquisitions or easements to permit a trail to continue east past Silver Lake Mountain over the long and tortuous cliff-faced ridge of Potter Mountain would make possible one of the wildest ridge walks in the northern Adirondacks.

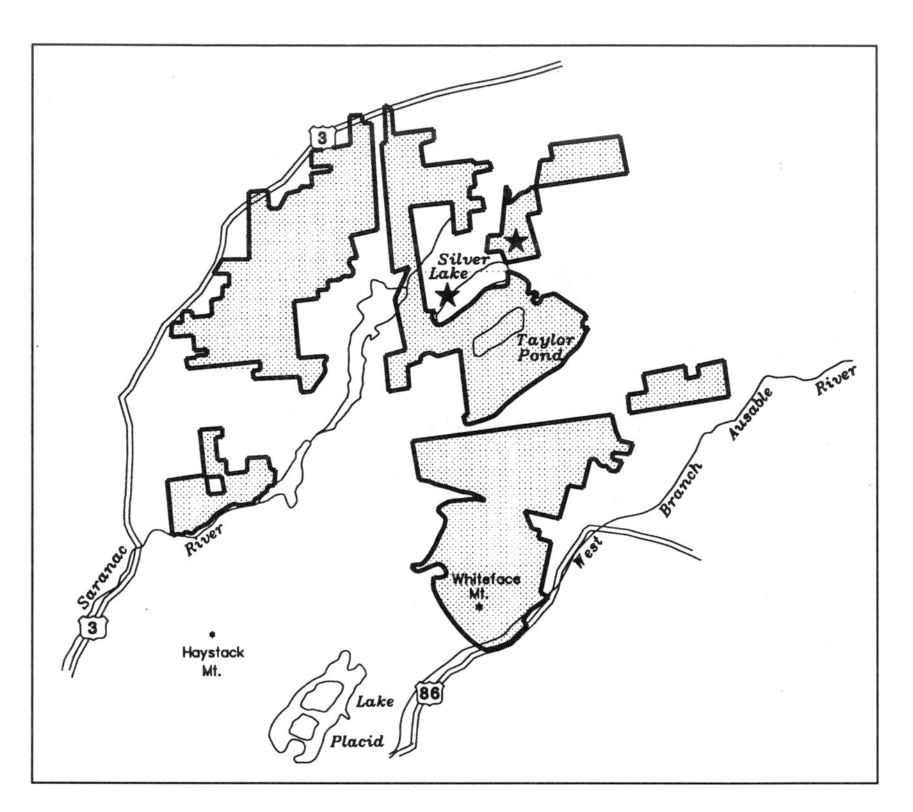

Views and Visits

The trail up Silver Lake is well marked. Walk as far as the ends of the boardwalk at Silver Lake Bog to sample its marvelous flora.

Catamount's rocky summit invites hikers to build cairns or stacks of small rocks. Esther and Whiteface Mountains form the distant horizon.

Photograph courtesy of Charles H. Bennett.

40 Small Northwestern Wild Forests

The Northwestern Adirondacks contains a few small patches of wild forest in a wide swath of private tracts that are among the most productive forest lands in the Park. The economics of the forest industry has forced many paper companies to question the viability of their Adirondack holdings and to consider granting easements to the state. These easements will open tracts to recreation, primarily hunting, and limit development, while the forests will still be harvested. Each easement is crafted individually, but almost all will open some land to the public, thus enhancing the recreational opportunities of the adjacent wild forests.

Add to these possibilities lands owned by Niagara Mohawk Power Corporation (NIMO) surrounding their Raquette River Reservoirs, of which Carry Falls is the largest. NIMO permits camping at specific sites along the chain of reservoirs and boating between the dams. NIMO has also helped build the fantastic Stone Valley Trails along the Raquette just outside the Blue Line. There, wooded trails follow the river through a series of rapids and splendid waterfalls. Put all these opportunities together with the state's wild forests and the result is a recreational reserve of many pieces where access to key scenic spots is possible, yet where forests can continue to produce the pulp and timber so vital to the economics of the north country.

Working to make the best recreational opportunities in this patchwork of state, private, and easement lands is not easy, but this area is a mosaic of different kinds of use that should be preserved.

GRASS RIVER WILD FOREST

The Grass River is known for its waterfalls. A patch of state land surrounding Lampson Falls on the Grass River already has lovely riverside trails. Upstream from the falls there is a stretch of flat water that invites canoeing. To the north, Harpers Falls on the North Branch of the Grass River is also on state land but needs a better marked trail. A string of waterfalls—Basford, Sinclair, Twin, Flat Rock, Rainbow, and Copper Rock—graces the South Branch of the Grass where it parallels Tooley Pond Road.

In 1998 the state announced a complicated purchase and easement agreement with Champion International that puts five of the falls and a long stretch of the Grass River in the Forest Preserve. The river has enormous possibilites for white-water canoeing. Now all that is needed are a series of short trails to picnic and fishing sites along the river and interpretive signs for the historic forge whose remains are near Twin Falls.

The easements leave 110,000 acres in private ownership. The public has access to most of these acres, and the agreements open up hundreds of miles of logging roads as a vast network of

Harpers Falls on the North Branch of the Grass River is only one of the scenic spots on the three branches of the Grass, but most of the other waterfalls are on private land.

Photograph courtesy of Wayne B. Virkler.

snowmobile trails. Like a chain of embroidery, these agreements seam together important private forests, leaving them in timber and pulp production, while making the highlights of the waterway gems of public access.

The long, flat, winding waterway along the upper South Branch of the Grass was already accessible by canoe or by the Nature Conservancy easement over the bed of the Grass River Railroad. With these important additions, the Grass, for many years a little-known river, is becoming one of the great recreation rivers of the Park.

Views and Visits

A very short trail leads to Lampson Falls, and the trails that follow the river downstream past a series of rapids, flumes, smaller falls, and chutes are a delight to walk.

WHITEHILL WILD FOREST

Sometimes called Clear Pond Wild Forest for its principal pond, this stretch of open peatland, spruce-tamarack swamps, and eskers was once the site of a scout camp. Old roads connecting several ponds beckon cross-country skiers.

RAQUETTE BOREAL WILD FOREST

The Lassiter Easement lands extend small tracts of state land near the Raquette River into a boreal reserve that includes the Jadwin Forest. Here old roads have become hiking and snowmobile trails through the deep, dark spruce-fir stands that typify the Adirondacks' most northern forest type.

Views and Visits

The Stone Valley trails near South Colton are just outside the Park boundary, but they are beautiful introductions to the majesty of the Raquette River.

41 Small Northern Wild Forests

State land patches at the northern edge of the park are so small they seem lost among tracts of private land. A few may grow, in fact should grow judiciously, in order to improve access; most will not grow in size because they are surrounded by roads and settlements. These patches are vital parts of the Park's quilt because of their special recreation features.

If you lump groups of these wild forests together, piecing them into larger management areas as the DEC does for technical reasons, their fragmented nature and zigzag boundaries emphasize the way they are too chopped up ever to be seen as wildernesses. If you look at each individually, observing its special features, it becomes obvious they are gems in the landscape, miniature wildernesses that deserve special attention.

A stark giant on Lake Kushqua. Photograph courtesy of the author.

BLOOMINGDALE BOG, VERMONTVILLE, KATE MOUNTAIN

A chain of small lots of state land stretches across these three areas. Bloomingdale Bog is a unique late-stage bog, dry enough now that it supports clumps of shrubs in mounds. Native orchids are secreted among the leatherleaf, pale rhododendron, and cranberries that thrive on hummocks of sphagnum moss. Tamarack and black spruce thrive in wet areas. Pines and balsams ascend drier upland sites. The raised bed of the abandoned railroad (the Chateaugay Division of the Delaware and Hudson Railroad) serves as a natural boardwalk through the bog.

Old roads offer ski and snowmobile trails north of Vermontville. Kate Mountain sits isolated amidst a plain with potato fields. The rock outcrops that offer such superb views are approached by bushwhacks—one from the north near the gorges around Negro Brook. Both the north and south approaches should be marked as trails with identified trailheads.

Separate chunks of state land extend north to Lake Kushqua, Rainbow Lake Narrows, and the lower part of the North Branch of the Saranac, protecting parts of a beautiful water route.

Views and Visits

Take a small canoe trip up the North Branch of the Saranac. Walk along the railroad through the bog.

42 Small Northeastern Wild Forests

Scattered in an arc across the northeastern Adirondacks are a number of Forest Preserve lots with peaks of various sizes. Some have access, some have trails; all could have both without filling in the patchwork of public and private lands that distinguishes the fairly settled regions of the northeastern Adirondacks. It is amazing how small some of these tracts of public land are; it is equally amazing to discover what some of them offer. It is wonderful to contemplate the blessings of having access and trails on all of them.

No great forests cover these tracts—almost all had fires. Many were initially stripped of their hardwoods for charcoal; others were logged for pulp. Some hills served as woodlots for the farms that surrounded them. None is without disturbance of some kind. They will never be joined together into large contiguous stretches of Forest Preserve, but they can be managed to provide recreation in a settled landscape.

Trails already exist on Split Rock, and plans are being formulated for extending them. Poke-o-moonshine has an excellent trail. The Four Ponds tract has canoe access trails. Gilligan has a trail with a fine view of Rocky Peak Ridge and beyond to the Dixes. A short trail to Elephant Head just outside the border of the Park has views north across Lake Titus.

Acquisitions or easements are needed for Ebenezer and Rattlesnake and Makomis. Easement accesses and trails will complete Norton Peak, Lyon Mountain, Alder Brook Mountains, Dannemora and Ellenburg, Daby, Jug, and the Haystack that is west of Norton Peak. Only a trail is needed to give access to the small hill west of Lincoln Pond; such access would greatly enhance that campground and that small tract of wild forest.

There has been a recent addition to the chain of small hills, although this one is owned by the Adirondack Conservancy. That organization has built a gem of a trail up the steep slopes of Coon Mountain. The town has provided trailhead parking. This exemplary cooperative effort shows methods that can be used to develop recreation throughout the northeastern Adirondacks.

Rattlesnake Mountain west of Willsboro (there is another west of the Northway) is privately owned, but a trail to its wonderful summit ridge is open to the public.

With the possibility of more than a dozen hikes for family outings on small mountains, all with excellent views, the northeastern Adirondacks has begun to offer alternative hikes to alleviate the overuse on the higher and more fragile peaks.

Views and Visits

The trails up Elephant Head, Coon, and Gilligan are all well marked. The trail up Coon Mountain is one of the best short hikes in the Adirondacks.

Vertical cliffs soar almost one hundred feet on the south side of Ebenezer Mountain.

Photograph courtesy of Wayne B. Virkler.

Brandt Lake church. Photograph courtesy of Charles H. Bennett.

Finishing the Quilt

Most quilts have a border of common fabrics that somehow finishes their designs. The Adirondack quilt has a border made up of private lands and working forests that completes the Park. Wider in the northwest and scattered throughout, these pieces are as vital to the economy of the region as the public lands are for tourism.

What is so exciting about our quilt is that the patches of public lands and working forests are sewn together with complex chains of stitches that represent roads and homes and businesses. The intersections are punctuated with fancy knots and bullions depicting charming settlements and hamlets. The embroidery adds richness to the Park that could never come from its wild lands alone.

But this book was planned to recognize the great variety of public lands within the Park. From all the descriptions, it is obvious that the public lands of our quilt are not of two primary colors. The Forest Preserve represents a wide spectrum of colors—a continuum of values where wilderness blends into recreation lands. Few wild forests lack wilderness characteristics. Those that do have generally been better served by trails. The most untouched wilderness areas lie in the southern Adirondacks and these include some of our designated wild forests, areas that have been almost forgotten by recreation planners.

Some of our finest wild lands are virtually inaccessible, separated from roads by private lands. A few wilderness patches are worn with multiple trails. But there is a real need to embroider a few more trails on some of the plainer pieces. The quilt needs constant care.

Despite these shortcomings, and others created by political decisions that separated our public lands into two basic categories, it is uplifting to view the Adirondack map as a magnificent quilt whose patterns represent a rich panoply of values and whose sum is the Adirondack Park.

Beaver Pond near Jones Hill.
Photograph courtesy of Charles H. Bennett.

BARBARA MCMARTIN has spent most of her life walking and exploring the Adirondacks. Her *Discover* series of guidebooks to the Adirondacks has given her the geographical basis for her environmental work and her historical books. She has served on the board or as an officer of several Adirondack advocacy groups and currently is chair of the Department of Environmental Conservation's Forest Preserve Advisory Committee. Her books include histories of the Park's hemlock bark tanning industry and of a pioneer family, the Benedicts, and *The Great Forest of the Adirondacks.*

Book design by Christopher Kuntze
Typeset in Adobe Minion and Albertus types
Printed by Friesens Press in Manitoba, Canada